AF784202

URETEROCELES

URETERECELES

Contemporary Diagnosis and Management

Edited by

MICHAEL LEONARD

401 Smyth Road, Ottawa, ON, K1H 8L1

LUIS GUERRA

401 Smyth Road, Ottawa, ON, K1H 8L1

ELSEVIER

ACADEMIC PRESS

An imprint of Elsevier

Academic Press is an imprint of Elsevier
125 London Wall, London EC2Y 5AS, United Kingdom
525 B Street, Suite 1650, San Diego, CA 92101, United States
50 Hampshire Street, 5th Floor, Cambridge, MA 02139, United States
The Boulevard, Langford Lane, Kidlington, Oxford OX5 1GB, United Kingdom

Library of Congress Cataloging-in-Publication Data
A catalog record for this book is available from the Library of Congress

British Library Cataloguing-in-Publication Data
A catalogue record for this book is available from the British Library

ISBN: 978-0-12-817888-1

For information on all Academic Press publications
visit our website at https://www.elsevier.com/books-and-journals

Publisher: Stacy Masucci
Editorial Project Manager: Samantha Allard
Production Project Manager: Maria Bernard
Designer: Christian Bilbow

Typeset by Thomson Digital

Contents

Contributors

Kourosh Afshar
Department of Urologic Sciences, Division of Pediatric Urology BC Children's Hospital, University of British Columbia, Vancouver, BC, Canada

Veridiana Andrioli
Department of Surgery, Division of Urology, Federal University of Sao Paulo, Sao Paulo, Brazil

Angela M. Arlen
Department of Urology, Yale University School of Medicine, New Haven, CT, United States

Paul F. Austin
Baylor College of Medicine, Texas Children's Hospital, Houston, TX, United States

Natalia Ballesteros
Division of Pediatric Urology, Nicklaus Children's Hospital;
Department of Urology, University of Miami, Miami, FL, United States

Jason P. Van Batavia
Pediatric Urology, The Children's Hospital of Philadelphia, Division of Urology, Perelman School of Medicine, University of Pennsylvania, Philadelphia, PA, United States

Brigitte Bonin
Department of Obstetrics and Gynaecology and Newborn Care, Division of Maternal-Fetal Medicine (MFM), Ottawa Hospital Research Institute (OHRI), The Ottawa Hospital (TOH), University of Ottawa, Ottawa, ON, Canada

Luis H. Braga
McMaster Children's Hospital, Department of Surgery, McMaster University, Hamilton, ON, Canada

Douglas A. Canning
Children's Hospital of Philadelphia; Perelman School of Medicine, University of Pennsylvania, Philadelphia, PA, United States

Yvonne Y. Chan
Division of Pediatric Urology, Ann and Robert H. Lurie Children's Hospital of Chicago, Chicago, IL, United States

Earl Y. Cheng
Division of Pediatric Urology, Ann and Robert H. Lurie Children's Hospital of Chicago, Chicago, IL, United States

Christopher S. Cooper
University of Iowa Hospitals and Clinics, Iowa City, IA, United States

Lisieux Eyer de Jesus
Pediatric Surgery and Urology, Antonio Pedro University Hospital; Servidores do Estado Federal Hospital, Rio de Janeiro, Brazil

Julia Beth Finkelstein
Pediatric Urology, Department of Urology, Boston Children's Hospital, Boston, MA, United States

Karen Fung-Kee-Fung
Department of Obstetrics and Gynaecology, Division of Maternal-Fetal Medicine (MFM), Ottawa Hospital Research Institute (OHRI), The Ottawa Hospital (TOH), University of Ottawa, Ottawa, ON, Canada

Rafael Gosalbez
Department of Urology, University of Miami, Miami, FL, United States

Luis Guerra
Children's Hospital of Eastern Ontario, University of Ottawa, Ottawa, ON, Canada

Melise A. Keays
Children's Hospital of Eastern Ontario, Department of Surgery, University of Ottawa, Ottawa, ON, Canada

Daniel Keefe
Children's Hospital of Eastern Ontario, Department of Surgery, University of Ottawa, Ottawa, ON, Canada

Michael P. Leonard
Children's Hospital of Eastern Ontario, University of Ottawa, Ottawa, ON, Canada

Armando J. Lorenzo
Pediatric Urology, Women's Auxiliary Chair in Urology and Regenerative Medicine, Hospital for Sick Children; Division of Urology, Department of Surgery, University of Toronto, Toronto, ON, Canada

Andrew E. MacNeily
Department of Urologic Sciences, Division of Pediatric Urology BC Children's Hospital, University of British Columbia, Vancouver, BC, Canada

Sameer Mittal
Pediatric Urology, The Children's Hospital of Philadelphia, Division of Urology, Perelman School of Medicine, University of Pennsylvania, Philadelphia, PA, United States

Felipe Moretti
Department of Obstetrics, Gynecology and Newborn Care, Division of Maternal-Fetal Medicine (MFM), The Ottawa Hospital, University of Ottawa, Ottawa, ON, Canada

João Luiz Pippi-Salle
Sidra Medical and Research Center, Doha, Qatar

Rodrigo Romao
Departments of Surgery and Urology IWK Health Centre, Dalhousie University, Halifax, NS, Canada

Ana Werlang
Department of Obstetrics, Gynecology and Newborn Care, Division of Maternal-Fetal Medicine (MFM), The Ottawa Hospital, University of Ottawa, Ottawa, ON, Canada

CHAPTER 1

Definition, classification, and embryology

Melise A. Keays[a], Daniel Keefe[a], Luis H. Braga[b]
[a]Children's Hospital of Eastern Ontario, Department of Surgery, University of Ottawa, Ottawa, ON, Canada
[b]McMaster Children's Hospital, Department of Surgery, McMaster University, Hamilton, ON, Canada

Historic descriptions

Ureterocele was first described by Lechler (Lechler, 1835), the lesion was described as a double bladder (ureterocele) found in a necropsy of a 3-month old infant. Bostroem later reviewed Lechler's description and concluded that his report was a ureterocele, and not a double bladder. He subsequently published a report of 3 cases of ureterocele discovered during necropsy, as well as 35 other ureterocele cases (Bostroem, 1884). The second account was by Lilienfeld in 1856 again described as a second bladder that projected into the lumen of the urethra (Lilienfeld, 1856). Several further reports followed with different descriptions, including inverted ureter (Caillé, 1888), cystic dilation of vesical end of ureter (Englisch, 1898), and ballooning of the ureter (Fenwick, 1903). The term ureterocele was first used by Stoeckel in 1907, where he described it as a ureteral hernia, or ureteralis vesicalis.

Incidence

At autopsy, the incidence of ureteroceles has been reported as 1 in 500 to 1 in 4000 cases (Decter, 1997; Uson, Lattimer, & Melicow, 1961). Ureteroceles occur 4-6 times more frequently in females than in males, and more commonly in Caucasians than other races (Shokeir & Nijman, 2002). There is no side predilection for ureteroceles and bilateral involvement has been found in 10% of cases (Coplen & Duckett, 1995).

Genetics

Ureteroceles have shown evidence of familial predominance in several cases. The rate of ureteral duplication in families is higher than in the general population. The first suggestion of a familial inheritance pattern dates back to 1939 when single system ureteroceles were described in identical twins (Riba, 1936).

Ureteroceles
http://dx.doi.org/10.1016/B978-0-12-817888-1.00001-X

Since then several other familial cases of ureteroceles have been described in both single and duplex systems (Abrams, Sutton, & Buchbinder, 1980; Ayalon, Shapiro, Rubin, & Schiller, 1979; Lachiewicz, Kogan, Levitt, & Weiner, 1985; Sozubir, Ewalt, Strand, & Baker, 2005). In some instances, there appears to be an autosomal dominant pattern of inheritance (Babcock, Belman, Shkolnik, & Ignatoff, 1977) although Mendelian patterns of inheritance are not generally observed (Schultza & Yoneka Todab, 2016). Family members of those with ureteroceles may exhibit varying phenotypes of congenital anomalies of the kidney and urinary tract (CAKUT). Current theories suggest that the specific gene mutations can manifest as different anomalies along the urinary tract.

Genetic mechanisms involved with ureterocele development have not been fully delineated and remain an area of ongoing study. Mechanistic theories suggesting genetic anomalies for ureterocele development have been primarily linked to embryologic development of the Wolffian duct and ureteric bud as well as the maturation and incorporation of the ureter into the bladder (Uetani & Bouchard, 2009). The primary genes involved with these embryologic steps include AGTR2 (angiotensin type-2 receptor), Foxc1/Foxc2 (murine forkhead/winged helix genes), BMP4 (bone morphogenic protein 4) and ret gene 17-20. The importance of these genes in ureteric development have led to focused studies to identify their potential as a source of CAKUT (Sozubir et al., 2005).

Despite the high correlation in ureterocele development in family members of those affected, screening has not been recommended. However, in twins of those with ureteroceles there is a higher correlation and it has been suggested they may benefit from ultrasound screening for ureteroceles (Sozubir et al., 2005).

Further genetic studies are required to describe specific genetic mechanisms of ureterocele development which may have implications across the care of those with ureteroceles, including diagnosis, management, and screening (Sozubir et al., 2005).

Embryology

Debate exists regarding the embryologic origin of ureteroceles, mainly due to the various types of ureteroceles, suspected to have different embryologic inciting events. Herein we review the process of normal embryologic development of the genitourinary system as well as the theories leading to the development of ureteroceles.

The mesonephric (Wolffian) duct (MD) becomes apparent by the 24th day of gestation and becomes the connection between the mesonephros and the cloaca. The MD gives rise to the ureteral bud (UB) by the 4th week of gestation, which develops cranially toward the metanephric blastema by the 5-6th week of gestation (Stephens, 1958). The UB eventually gives rise to the ureter, renal pelvis, calyces, papilla, and collecting ducts. Urinary tract development involves reciprocal induction of the UB with the metanephros, which will develop into proximal nephron structures.

The distal MD is known as the common excretory duct and progresses anteromedially toward the primitive cloaca and fuses ventrally by the 4th week of gestation. At the 7th week of gestation, the ventral cloaca becomes the urogenital (UG) sinus and the dorsal portion becomes the rectum. When the right and left common excretory ducts fuse in the midline this forms the precursor to the bladder trigone (Brookes & Zietman, 1998). At 8 weeks of gestation the common excretory duct and UB enter the posterior aspect of the UG sinus. The nephric duct and UB undergo apoptosis and dissociate allowing caudal migration of the nephric duct and lateral and cranial migration of the ureteric orifice. By the 12th week of gestation, both orifices have reached their final locations with the MD located in the posterior urethra.

UB incorporation into the UG sinus is an integral part to normal urinary system development. When the UB originates lower than normal on the MD there is early incorporation into the UG sinus leading a more lateral and cranially migrated ureter. This predisposes the patient to vesicoureteral reflux. If the UB originates higher on the MD it becomes incorporated into UG sinus later than usual leading to abnormal ureteric orifice locations ranging from minor displacement toward the bladder neck or complete failure to incorporate into the bladder altogether leading to ectopic ureter insertions in the urethra or mesonephric remnant structures. In addition, development of two UBs, or an early bifurcating UB, can lead to duplicated systems which can affect the timing and position of UB incorporation into the UG sinus thus leading to anomalous urinary tracts (Ambrose & Nicolson, 1964). When a duplicated system exists, Weigert-Meyer rule explains that the medial and caudal orifice is associated with the upper pole of the kidney. In general, this ureter is more frequently associated with ureterocele and obstruction.

The exact embryologic mechanism for ureterocele development is not fully known. There are several proposed mechanisms. Chwalla described a 2-cell layer membrane present at the time of UB formation from the MD. This membrane breaks down to facilitate patency of the ureter. The incomplete breakdown of Chwalla's membrane between the UB and the MD results in obstruction that leads to formation of ureterocele (Chwalle, 1927). Criticisms of this theory are that this explains the stenotic type of

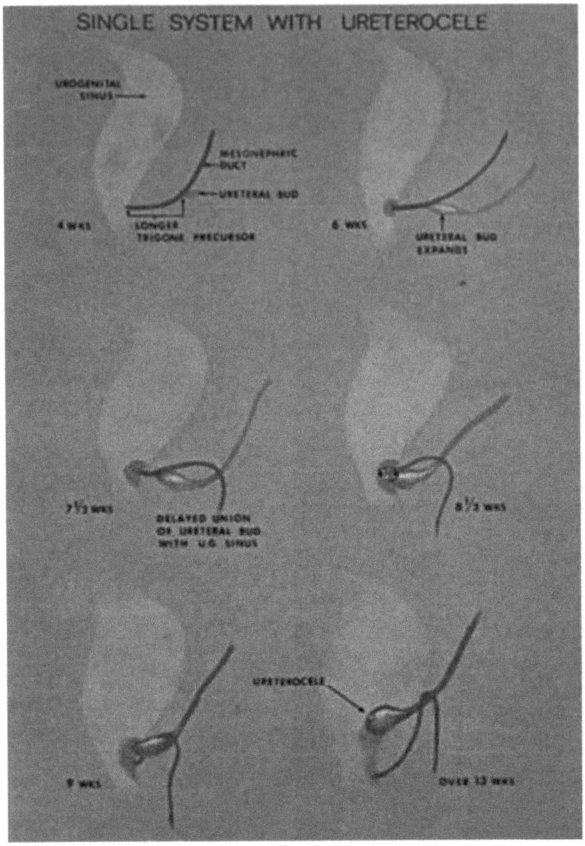

Figure 1.1 ***Ureterocele in single system, possibly due to higher location of bud on mesonephric duct and delayed luminal union of bud and duct; ureteral dilatation initiated while ureter still attached to mesonephric duct; fully formed ureterocele would finally meet urogenital sinus and ascend.*** *Borrowed (with permission) from Tanagho, 1976.*

ureterocele formation however many ureteroceles have patulous ureteral orifices. Tanagho described another mechanism which stems from delayed establishment of the UB lumen resulting in ureteral expansion (Tanagho, 1976)(Figs. 1.1–1.3). Lastly, Stephens postulated from histologic studies that there exists intrinsic muscle deficiency which affects bladder trigone development allowing for ureterocele formation (Stephens, 1958).

Classification

Several different ureterocele classifications have been described since their initial reporting. The current and most widely used **classification system has been proposed in 1984 by the American Academy of Pediatrics** (Glassberg et al., 1984).

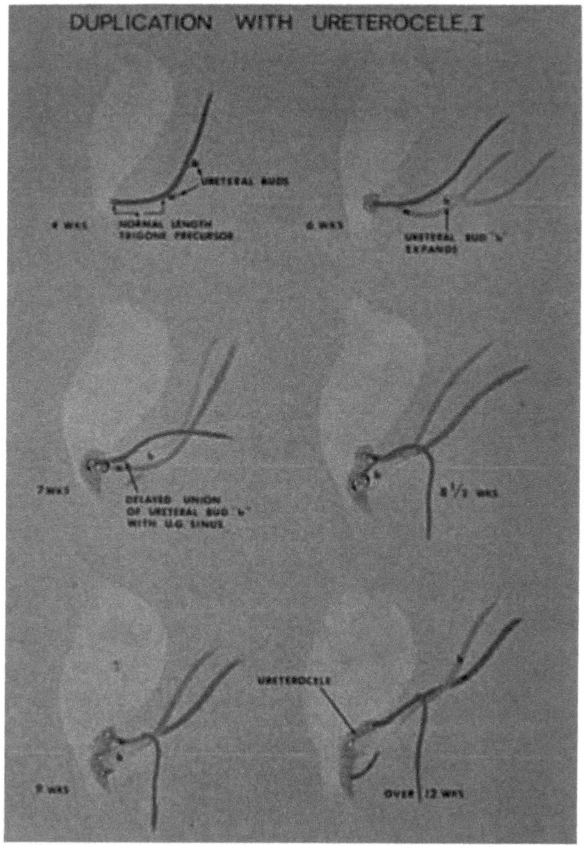

Figure 1.2 ***Ureterocele in duplicated system (always involving higher bud).*** If bud (a) is in normal position, it will end up at normal site on bladder base; bud (b) is higher, starts its cystic dilatation while still attached to mesonephric duct and will end up as small ureterocele at lower level. *Borrowed (with permission) from Tanagho, 1976.*

Location: Intravesical vs. Ectopic

The AAP classification distinguishes intravesical ureteroceles which are entirely within the bladder compared to ectopic ureteroceles, where some portion is situated permanently at the bladder neck or in the urethra. The committee proposed the avoidance of the term orthotopic or simple as these terms have led to confusion. **Note, an intravesical ureterocele that is obstructing the bladder neck by prolapse would still be considered an intravesical ureterocele** (Figs. 1.4 and 1.5).

Orifice: Stenotic, Sphincteric, Sphincterostenotic or Cecoureterocele

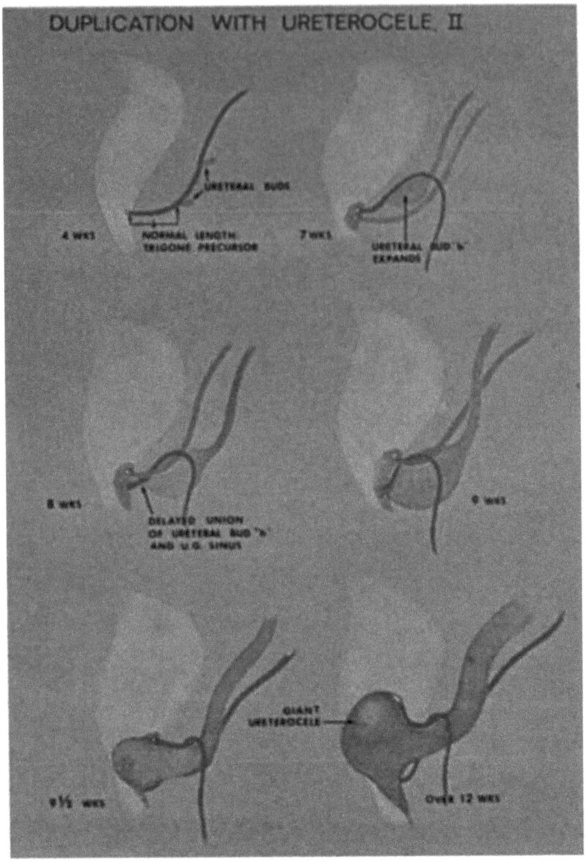

Figure 1.3 ***Duplication with giant ureterocele involving higher bud.*** If bud (b) is too high on mesonephric duct, more time is available for cystic expansion of its terminal part, which will join urogenital sinus late and end up in location lower than normal (either above internal meatus or ectopic, below level of internal meatus) *Borrowed (with permission) from Tanagho, 1976.*

The subcommittee refers that further sub-classification, as proposed by Stephens (F. Stephens, 1968) in 1968 can further describe and characterize ureteroceles as:

Stenotic: A ureterocele caused by an intrinsic stenosis of the ureteric orifice. The position may be normal or ectopic according to Stephens, but is normally located on the lateral aspect of the bladder base. The ureteric orifice lies on the dome of the bladder wall protrusion and the grade of stenosis may range from almost normal to pinpoint in caliber (Fig. 1.6A).

Sphincteric: A normal or enlarged orifice or ureter in the floor of the urethra constricted by the internal sphincteric muscle fibers of the urethra.

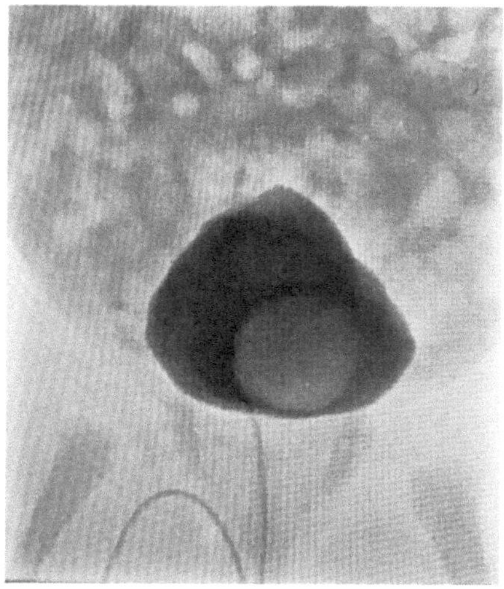

Figure 1.4 ***Intravesical ureterocele appears as filling defect on VCUG.***

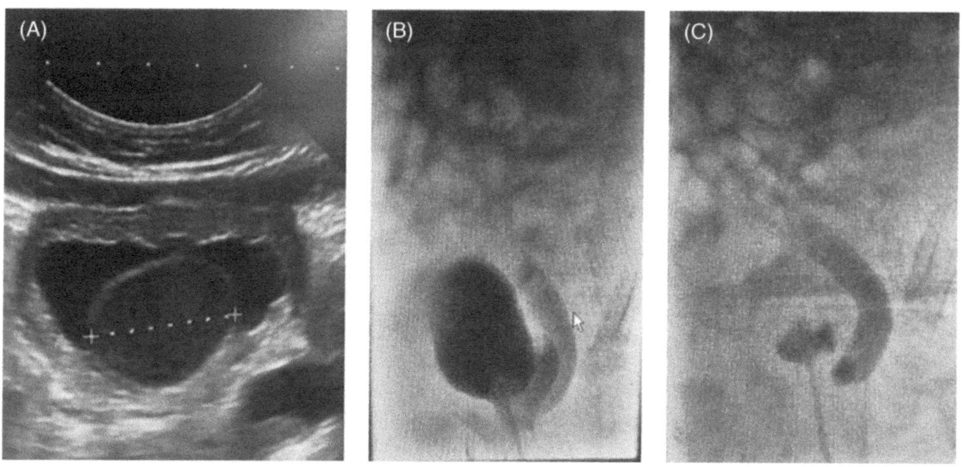

Figure 1.5 Ectopic ureterocele seen on ultrasound may have similar appearance as an intravesical ureterocele (A). On voiding phase of VCUG can see reflux from ectopic orifice (B). Post-void images demonstrate residual contract in left ureter (C).

The orifice is ectopic and lies on the posterior wall of the urethra in the internal sphincter zone. The continuous squeezing action of the internal sphincter during bladder filling is responsible for the obstructing mechanism in this type. The orifice itself does not need to be obstructive, it may be larger than normal, readily admitting the beak of the cystoscope, but it is completely blocked off by the compression of the internal sphincter muscle of the urethra during bladder filling. It becomes unblocked with voiding,

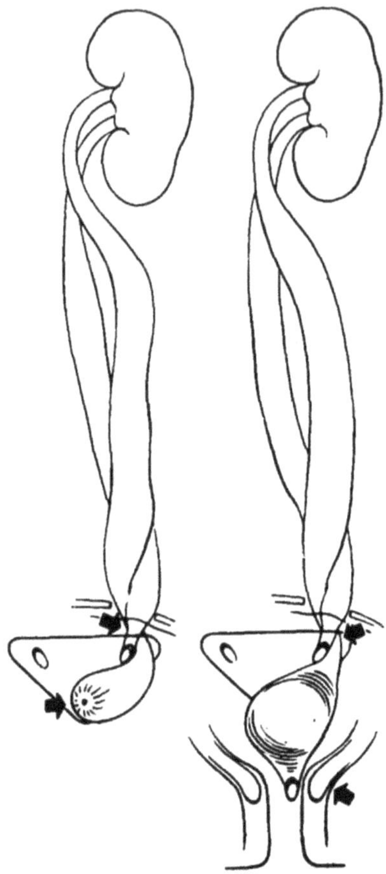

Figure 1.6 ***Stenotic and sphineteric ureteroceles (left and right, respectively) situated on ectopic ureters.*** The *lower arrows* indicate the sites of the obstructions which cause the ureteroceles. The *upper arrows* show the points at which the turgid ectopic ureters and ureteroceles obstruct the orthotopic ureters and their orifices. *Borrowed (with permission) from Stephens, 1958.*

with sphincteric relaxation. Just like the bladder, the sphincteric ureterocele fills with drainage of the affected renal moiety and only empties with voiding. These ureteroceles tend to be quite large (Fig. 1.6B).

Historically, ectopic ureteroceles that inserted beyond the bladder neck into the urethra were called cecoureteroceles. Cecoureteroceles occur when the orifice of the affected ureter is within the bladder, but the cavity of the ureterocele extends beyond the bladder neck into the urethra, posing some challenges for endoscopic incision. In some cases, just puncturing the ureterocele base in the bladder is not enough, as urine may fill the distal cavity of the ureterocele, causing outlet obstruction due to a wind sock effect.

These images (Fig. 1.7) represent a infant female right sided upper pole ectopic ureterocele with sphincteric opening. In the first image, you can see dilation of the upper moiety. In the second image, you can visualize a dilater upper pole ureter that connects to a ureterocele. In the third bladder image, a small thin walled ureterocele appears.

Fig. 1.8 portrays VCUG images from the same patient demonstrating a filling defect at the base of the bladder. On the voiding phase on static images, the ureterocele defect appears to decompress as the ureterocele is able to empty with relaxation of the external urethral sphincter.

Description of Fig. 1.9. In these series of images within the female urethra you can see that in image A at the level of the external sphincter, the urethra does not have a normal appearance and there is a clear area of elevation / extra tissue at the 6-8 o'clock position (patient's right side) which corresponds with the ultrasound findings of a right upper pole ureterocele. When the scope is pulled back slightly, with irrigation fluid slowly running, a patulous opening at the level of the sphincter is found. This opening is large enough to accommodate the rigid cystoscope and in the 3rd image, you can see inside the ureterocele cavity. The final image is just inside the bladder neck where you can see the ureterocele protruding into bladder.

Sphinctero- stenotic: A combination of the stenotic and sphincteric subtypes.

Collecting system: Single vs. Duplex

Ureteroceles are associated with single system (single ureter) vs. duplicated collecting systems with two complete ureters. The German pathologist Carl Weigert first recognized that in complete ureteral duplication, the ureter with a more medial and caudal orifice (known as the ectopic ureter) reaches the upper renal moiety (Weigert, 1877). The other ureter with a more lateral and cephalad orifice, connects to the lower moiety. Later, Meyer (Meyer, 1946) recognized that this disposition is almost universal in cases of complete ureteral duplication and has become known as the Weigert-Meyer Law.

In duplex systems, the ureterocele is associated with the upper pole moiety and tends to be obstructed and the lower pole tends to have vesicoureteral reflux. The classic sign of this finding on VCUG is called the drooping lily sign (described in Fig. 1.10), which shows reflux into the lower renal moiety that is displaced downward due to the severely dilated and obstructed upper pole.

Fig. 1.10 shows right duplex system with upper pole intravesical ureterocele and lower refluxing moiety. On the VCUG image on the left, a "drooping lilly" sign of lower pole VUR with no calyces pointing upwards at the top of the renal pelvis. Ultrasound images demonstrate upper pole

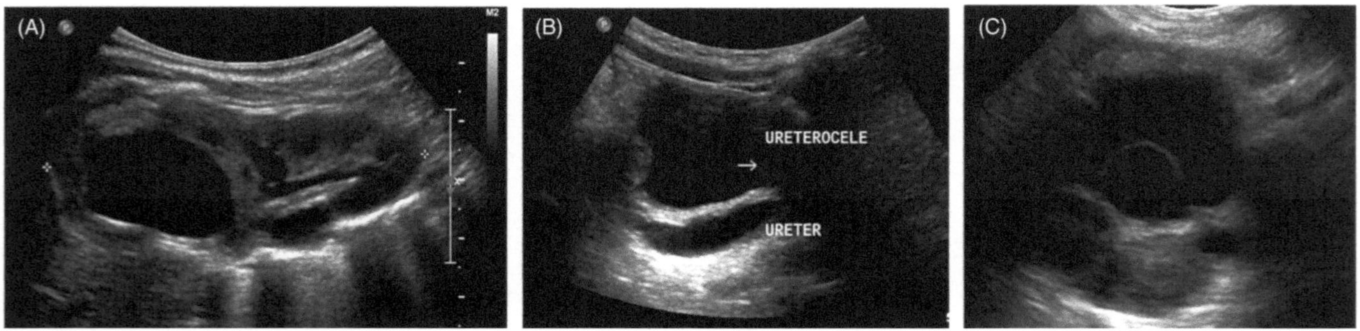

Figure 1.7 ***Ultrasound images of duplex right ectopic ureterocele with sphincteric opening.*** (A) Demonstrates upper pole dilation. (B) Ureter enters ureterocele with ureteral dilation appreciated. (C) Smooth thin walled ureterocele visible in bladder.

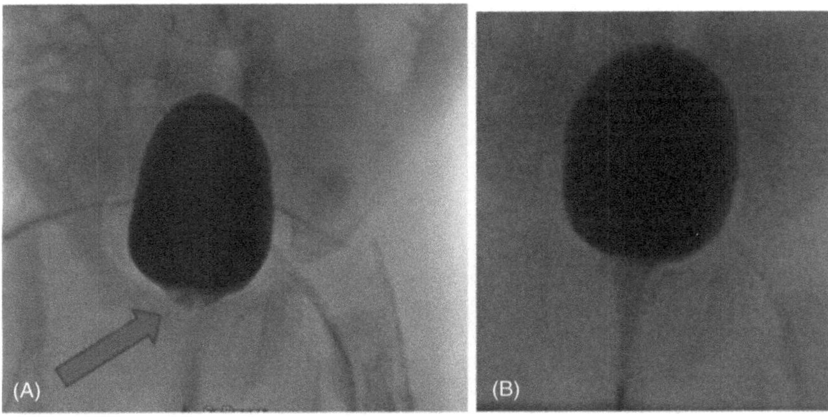

Figure 1.8 ***VCUG of female with ectopic ureterocele with sphincteric opening.*** (A) Filling defect noted to protrude into bladder neck. (B) During live fluoroscopy, the filling defect can become less apparent during voiding phase when ureterocele can decompress due to relaxation of sphincter.

hydronephrosis with thinning. The surgical view of a bladder depicts an intravesical ureterocele, the upper pole ureteral orifice coming out of the ureterocele is cannulated with a feeding tube.

Approximately 80% of ureteroceles are associated with the upper pole of a duplex collecting system, and 60% of these are ectopic (Coplen & Duckett, 1995), whereas intravesical ureteroceles are more common in single systems.

Functional classification system

Churchill (Churchill, Sheldon, & McLorie, 1992) proposed a functional classification of ectopic ureteroceles based on the amount of functional renal tissue, or renal units at risk of damage from obstruction or high-grade reflux (Table 1.1).

Ureterocele disproportion is seen in cases that appear to have a single system, however cystoscopic evaluation will demonstrate two ureteral orifices, the upper ureter is not dilated and tortuous, as usually seen in classic ectopic ureteroceles of duplicated systems, and leads to a completely dysplastic moiety. A classic ureterocele is seen in the bladder, but the ipsilateral kidney appears completely normal.

Dilated ectopic ureter mimicking a ureterocele

Rarely an ectopic dilated ureter may appear like ureterocele. The lower dilated end of a megaureter or dilated ectopic ureter may elevate the trigone resembling the protrusion caused by the ureterocele in the bladder. This clinical scenario can sometimes pose diagnostic challenges on ultrasound and it is often referred to as pseudo ureterocele (Fig. 1.11A).

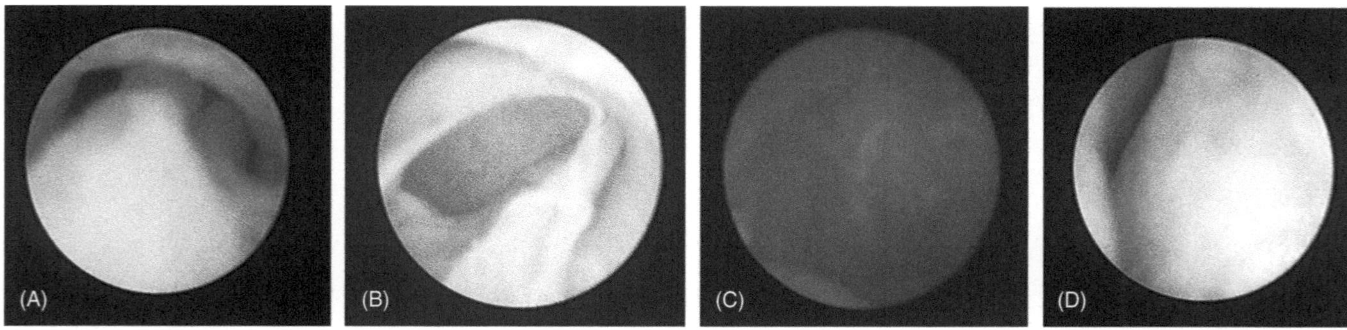

Figure 1.9 ***Cystoscopic views of sphincteric (non-stenotic) ureterocele.*** (A) Sphincter view, (B) patulous ureterocele opening located at level of external urethral sphincter, (C) view from within the ureterocele, and (D) view just inside bladder neck with ureterocele seen inside bladder.

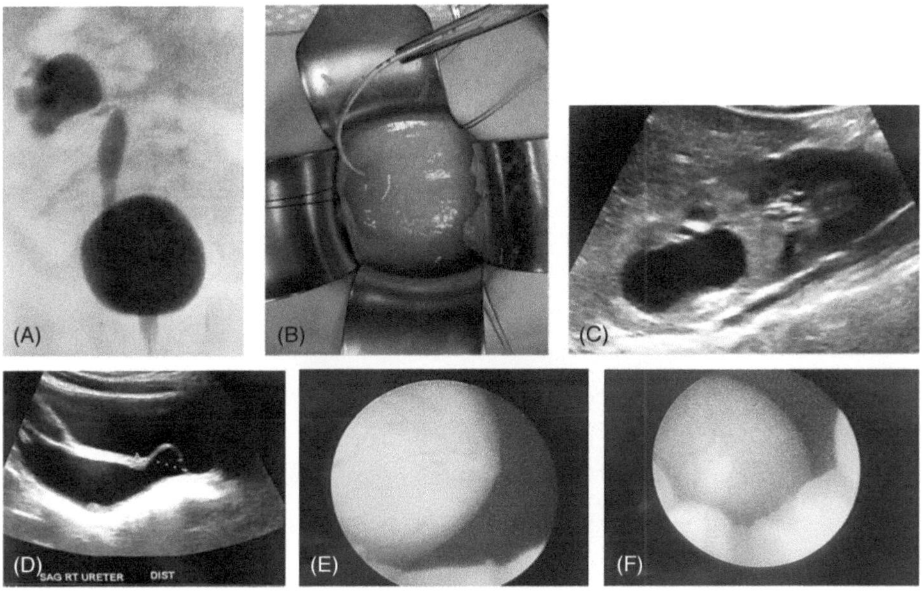

Figure 1.10 ***Right duplex system with upper pole intravesical ureterocele.*** (A) VCUG images demonstrate classic drooping lilly sign. (B) Open surgical view demonstrates intact ureterocele that is cannulated by feeding tube. (C) US demonstrating upper pole hydronephrosis. (D) Bladder view demonstrating ureterocele in bladder. (E) Cystoscopic view of intravesical ureterocele. (F) Cystoscopic view of completely intravesical ureterocele from bladder neck.

Table 1.1 Functional classification of ectopic ureteroceles.

Grade	Definition	Significance
I	One renal unit in jeopardy	Only the upper pole parenchyma subtended by the ureterocele demonstrates significant injury. Other renal units are normal, minimally hydronephrotic, and have low grade (I or II) reflux.
II	Entire ipsilateral kidney in jeopardy	As in grade I. In addition, the lower pole ipsilateral to the ureterocele is significantly hydronephrotic or associated with high-grade (Ill, IV,V) reflux. Contralateral kidney is normal or has mild hydronephrosis or reflux.
III	All renal units in jeopardy	As in grade II. In addition, contralateral kidney significantly hydronephtotic or associated with high-grade reflux.

Borrowed (with permission) from Churchill et al., 1992.

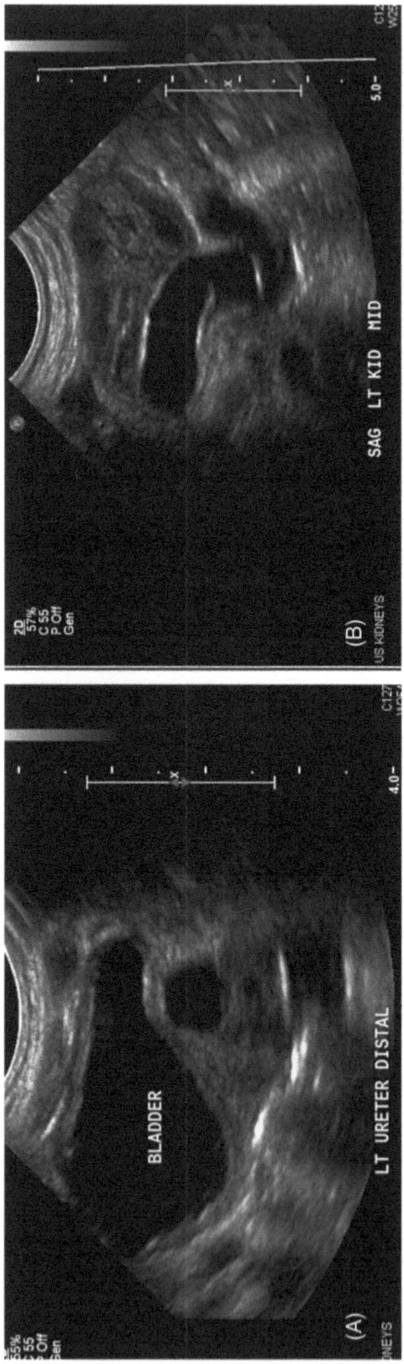

Figure 1.11 ***Dilated ectopic ureter mimicking a ureterocele.*** (A) Thick walled lining in bladder gives clue that this is an ectopic ureter rather than ureterocele. (B) Upper dilation on US with associated hydroureter.

Cystoscopic evaluation can provide insights into the type of ureterocele encountered. During cystoscopy, low fluid filling of the bladder in the early phase, as well as knowledge of normal trigonal anatomy and typical landing zones for ureteral orifices may help reorient during a difficult case. Bladder emptying with slow refilling may help revisualization of a decompressed ureterocele. If the orifice is not visible within the bladder, low volume irrigation below bladder neck may help visualize the opening of an ectopic ureter.

The subsequent chapters cover the management and outcomes of both conservative and operative approach to pediatric ureteroceles.

References

Abrams, H. J., Sutton, A. P., & Buchbinder, M. I. (1980). Ureteroceles in siblings. *The Journal of Urology, 124*(1), 135–1135 doi:10.1016/s0022-5347(17)55329-x.

Ambrose, S. S., & Nicolson, W. P. (1964). Ureteral reflux in duplicated ureters. *The Journal of Urology, 92*(5), 439–443 doi:10.1016/S0022-5347(17)63983-1.

Ayalon, A., Shapiro, A., Rubin, S. Z., & Schiller, M. (1979). Ureterocele—a familial congenital anomaly. *Urology, 13*(5), 551–553 doi:10.1016/0090-4295(79)90472-2.

Babcock, J. R., Belman, A. B., Shkolnik, A., & Ignatoff, J. (1977). Familial ureteral duplication and ureterocele. *Urology, 9*(3), 345–349.

Bostroem, E. (1884). *Beitrage zur pathologischen Anatomie der Nieren, Heft I.* J. C. B. Mohr: Freiburg und Tubingen.

Brookes, M., & Zietman, A. (1998). *Clinical embryology: a color atlas and text.* Boca Raton: CRC Press.

Caillé, A. (1888). Prolapse of the inverted lower portion of the right ureter through the urethra in a child two weeks old. *The American Journal of the Medical Sciences, 95*(5), 481 1827-1924.

Churchill, B. M., Sheldon, C. A., & McLorie, G. A. (1992). The ectopic ureterocele: a proposed practical classification based on renal unit jeopardy. *Journal of Pediatric Surgery, 27*(4), 497–500 doi:10.1016/0022-3468(92)90346-9.

Chwalle, R. (1927). The process of formation of cystic dilatation of the vesical end of the ureter and of diverticula at the ureteral ostium. *The Urologic and Cutaneous Review, 31*, 499–504.

Coplen, D. E., & Duckett, J. W. (1995). The modern approach to ureteroceles. *The Journal of Urology, 153*(1), 166–171 doi:10.1097/00005392-199501000-00068.

Decter, R. M. (1997). Renal duplication and fusion anomalies. *Pediatric Clinics of North America, 44*(5), 1323–1341 doi:10.1016/S0031-3955(05)70559-9.

Englisch, J. (1898). Ueber cystenartige Erweiterung des Blasenendes des Harnleiters. *Centralbl. fd Krankh. d. Harn-u. Sex. -Org, 9*, 373–406.

Fenwick, E.H. (1903). The value of ureteric meatoscopy in obscure diseases of the kidney: a study in clinical and operative surgery. London: J.&A. Churchill.

Glassberg, K. I., Braren, V., Duckett, J. W., Jacobs, E. C., King, L. R., Lebowitz, R. L., Classification, S. o. U., American Academy of Pediatrics. (1984). Suggested terminology for duplex systems, ectopic ureters and ureteroceles. The Journal of Urology, 132(6), 1153-1154. doi:10.1016/s0022-5347(17)50072-5.

Lachiewicz, A. M., Kogan, S. J., Levitt, S. B., & Weiner, R. L. (1985). Concurrent agenesis of the corpus callosum and ureteroceles in siblings. *Pediatrics, 75*(5), 904–907.

Lechler (1835). Mediz. Korrespondenzblatt d. Wurtemberg. *Arztl. Vereins*, *4*, 23.
Lilienfeld, B. (1856). Beitrage zur Morphologie und Entwicklungsgeschichte der Geschlechtsorgane und Beschreibung einer interessanten Missbildung. Marburg.
Meyer, R. (1946). Normal and abnormal development of the ureter in the human embryo-a mechanistic consideration. *The Anatomical Record*, *96*(4), 355–371.
Riba, L. W. (1936). Ureterocele: with case reports of bilateral ureterocele in identical twins. *British Journal of Urology*, *8*(2), 119–131 doi:10.1111/j.1464-410X.
Schultza, K., & Yoneka Todab, L. (2016). Genetic basis of ureterocele. *Current Genomics*, *17*(1), 62–69 doi:10.2174/1389202916666151014222815.
Shokeir, A., & Nijman, R. (2002). Ureterocele: an ongoing challenge in infancy and childhood. *BJU International*, *90*(8), 777–783 doi:10.1046/j.1464-410X. 2002.02998.x.
Sozubir, S., Ewalt, D., Strand, W., & Baker, L. A. (2005). Familial ureteroceles: an evidence for genetic background? *Turkish Journal of Pediatrics*, *47*(3), 255–260.
Stephens, F. (1968). Aetiology of ureteroceles and effects of ureteroceles on the urethra. *British Journal of Urology*, *40*(4), 483–1483.
Stephens, F. D. (1958). Ureterocele in infants and children. *Australian and New Zealand Journal of Surgery*, *27*(4), 288–295.
Tanagho, E. A. (1976). Embryologic basis for lower ureteral anomalies: a hypothesis. *Urology*, *7*(5), 451–464 doi:10.1016/0090-4295(76)90179-5.
Uetani, N., & Bouchard, M. (2009). Plumbing in the embryo: developmental defects of the urinary tracts. *Clinical Genetics*, *75*(4), 307–317 doi:10.1111/j.1399-0004.2009.01175.x.
Uson, A. C., Lattimer, J. K., & Melicow, M. M. (1961). Ureteroceles in infants and children: a report based on 44 cases. *Pediatrics*, *27*(6), 971–983.
Weigert, C. (1877). Ueber einige bildungsfehler der ureteren. *Virchows Archiv*, *70*(4), 490–501.

CHAPTER 2

Prenatal diagnosis of ureteroceles

Ana Werlang[a], Brigitte Bonin[b], Felipe Moretti[a], Karen Fung-Kee-Fung[c]
[a]Department of Obstetrics, Gynecology and Newborn Care, Division of Maternal-Fetal Medicine (MFM), The Ottawa Hospital, University of Ottawa, Ottawa, ON, Canada
[b]Department of Obstetrics and Gynaecology and Newborn Care, Division of Maternal-Fetal Medicine (MFM), Ottawa Hospital Research Institute (OHRI), The Ottawa Hospital (TOH), University of Ottawa, Ottawa, ON, Canada
[c]Department of Obstetrics and Gynaecology, Division of Maternal-Fetal Medicine (MFM), Ottawa Hospital Research Institute (OHRI), The Ottawa Hospital (TOH), University of Ottawa, Ottawa, ON, Canada

Introduction

The ability to view the fetus in its intrauterine environment with ultrasound imaging has revolutionized obstetric care and prompted the development of the concept of "the fetus as a patient," enabling antenatal fetal diagnoses and direct fetal therapy. In developed countries, the standard of obstetric care incorporates sonographic fetal anatomic survey into routine antenatal care. Insights into development and natural history of renal tract anomalies have been facilitated by technological developments in obstetric ultrasound imaging and the incorporation of routine ultrasound imaging protocols into structural evaluation of every fetus. Published clinical practice guidelines for obstetric imaging specialties provide direction for imaging standards of the fetal renal tract (AIUM–ACR–ACOG–SMFM–SRU Practice Parameter for the Performance of Standard Diagnostic Obstetric Ultrasound Examinations., 2018; International Society of Ultrasound in Obstetrics and Gynecology (ISUOG) practice guidelines: Performance of first-trimester fetal ultrasound scan, 2013). Formerly limited to examination of the fetus in the second and third trimesters, obstetric imaging of the fetus has been extended into the first trimester as an integral part of established aneuploidy screening programs (International Society of Ultrasound in Obstetrics and Gynecology (ISUOG) practice guidelines: Performance of first-trimester fetal ultrasound scan, 2013). First trimester imaging of the fetus has provided the opportunity for early structural survey of the fetus. This has enabled the early detection of fetal renal anomalies, informed parental counselling around fetal renal diagnoses and facilitated early, antenatal referral of affected fetuses and their parents to Maternal Fetal Medicine specialists and Pediatric Urologists for consideration of fetal or neonatal therapy. This chapter will chronicle the

Ureteroceles
http://dx.doi.org/10.1016/B978-0-12-817888-1.00002-1

development of the fetal urinary tract viewed sonographically, outline the imaging features characterizing antenatal diagnosis of ureteroceles, highlight the differential diagnosis of this condition, and provide insights into in utero options for prenatal management.

Prenatal sonoembryology of the urinary tract

Kidneys

Development of the urinary tract during fetal life can be chronicled by ultrasound. Production of a hypotonic urine characteristic of the fetus starts with development of the kidneys, first visible on ultrasound imaging at approximately 10 weeks gestational age. Fetal kidneys display a "bean-like" configuration on para-spinal views of the fetal abdomen in the sagittal, transverse, and coronal imaging planes (Fig. 2.1A and B). The presence of fetal kidneys in their usual intra-abdominal location can be confirmed on color Doppler imaging by visualization of bilateral renal arteries branching from the abdominal aorta (Fig. 2.1C).

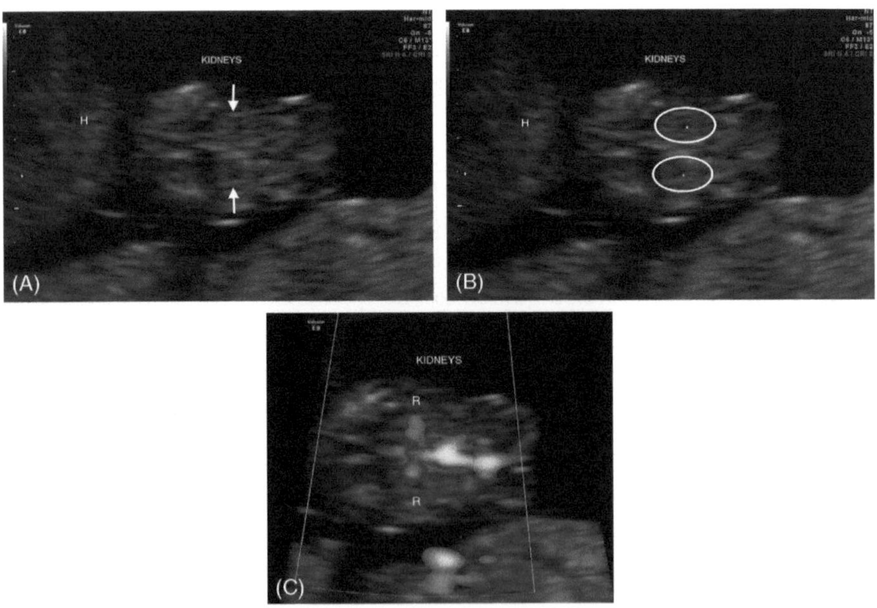

Figure 2.1 (A) Kidneys in coronal view, 13 weeks GA. The kidneys are hyperechogenic structures *(arrow)* with hypoechoic pelvis *(star)* sitting along the lumbar spine; (B) Kidneys *(ellipses)* and renal pelvis *(stars)* are outlined; (C) renal arteries can be appreciated bilaterally (R). *H*, fetal head.

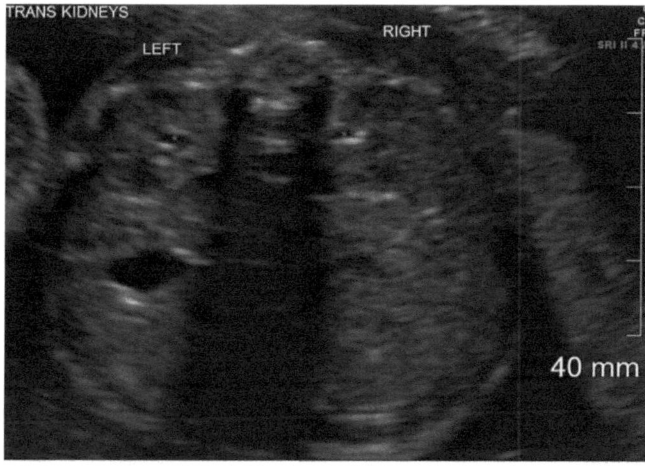

Figure 2.2 ***Transverse view of normal kidneys, 25 weeks.*** The renal parenchyma is less echogenic and more homogeneous. The anteroposterior diameter of renal pelvises *(stars)* are measured in this plane.

The fetal kidneys should be visible in the renal fossae at the time of the second trimester fetal anatomic survey (Fig. 2.2). If only identified unilaterally, or if only one renal artery can be visualized on color Doppler imaging then an anomaly of kidney number (e.g., renal agenesis, Fig. 2.3), or abnormality of renal position, or renal dysmorphology (e.g., horseshoe kidney) is likely. Growth of the kidney continues to term. Various nomograms for fetal renal lengths have been published (Darahem Tedesco et al., 2009; Shin,

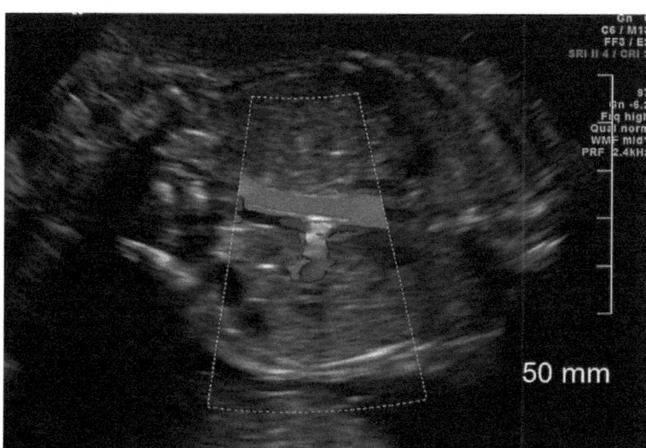

Figure 2.3 ***Unilateral renal agenesis, 26 weeks GA.*** *Note*: Single renal artery on Color Doppler on coronal view.

Seo, Kim, & Park, 2007; van Vuuren et al., 2012; Yu, Chang, Chang, Ko, & Chen, 2000). The ratio of fetal renal circumference to abdominal circumference remains stable over gestation (Mileto et al., 2018).

Echogenicity of the kidney on ultrasound imaging exhibits variation over gestation. Computerized-based numerical pixel analysis of digital images of the fetal kidneys throughout gestation has concluded that in early pregnancy the fetal kidneys appear more echogenic than in late gestation (Hershkovitz, Amichay, & Stein, 2011). In contrast to the echogenicity of the fetal liver, which remains fairly constant throughout gestation, by late pregnancy the fetal kidneys are much more homogeneous in appearance in mid-pregnancy.

Differentiation of the cortex from the medullary region of the fetal kidney is possible in the second trimester. In the third trimester the normal fetal renal cortex becomes thicker and evolves to a more hypoechoic echogenicity compared to the fetal liver and spleen, while the corticomedullary ratio (CMT) decreases as the medulla increases in relative size (Fig. 2.4A and B) (Devriendt, Cassart, Massez, Donner, & Avni, 2013). Failure to visualize a decrease in this ratio in the third trimester or a renal cortex that appears echogenic in comparison to the fetal liver or spleen is suspicious for the presence of nephropathy.

Routine anatomic survey for anomalies in the second trimester of pregnancy includes both sagittal and transverse views of the fetal kidneys. The renal pelvis is easily demonstrated on these latter views and diameters of less than 4 mm in antero-posterior (AP) in the second trimester are considered to be in the normal range. Diameters greater than 7 mm in the third trimester are generally considered criteria for post-natal evaluation, which can include imaging and referral to Pediatric Urology, particularly if greater than 15 mm. Dilatation of the fetal renal tract can be a marker

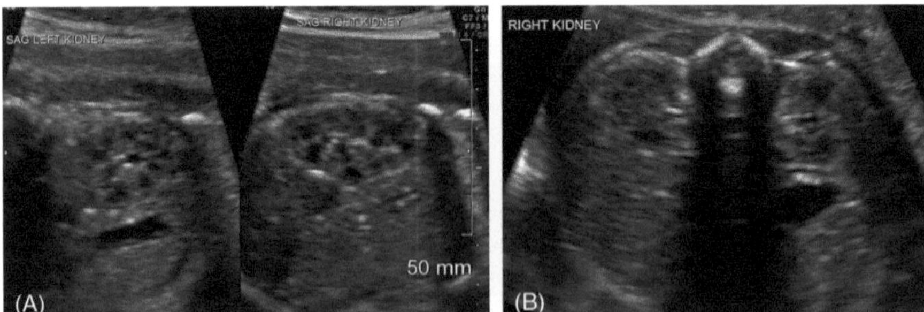

Figure 2.4 Sagittal (A) and transverse; (B) views of normal kidneys, 31 weeks. *Note:* The hyperechogenic parenchyma and hypoechogenic renal medullae.

of antenatal hydronephrosis, one of the most commonly diagnosed prenatal ultrasound findings, affecting 1%–5% of pregnancies (Chapman, 2012). However, it is not exclusive to it and it can be observed as a transient finding in other obstetric conditions that have polyuria as a feature, such as gestational diabetes, Twin-Twin Transfusion Syndrome, and polyhydramnios. While there is general agreement that enlargement of the renal pelvis, or pyelectasis, can be an important sonographic marker of an obstructive uropathy, various grading systems have been suggested to improve antenatal detection in an attempt to limit long-term urological morbidity that can be associated with late diagnosis of this condition (Chalmers, Meyers, Brodie, Palmer, & Campbell, 2016; Fernbach, Maizels, & Conway, 1993; Gramellini et al., 2006). Besides rating the degree of pyelectasis by assessment of the renal pelvis AP diameter these grading systems incorporate other sonographic features of the renal tract into their scoring systems such as calyectasis, parenchymal thinning, ureteric dilatation, and presence of bladder abnormalities (Nguyen et al., 2014).

Ureters

Almost invariably, visualization of the fetal ureters at any time in pregnancy is a pathological finding and warrants further investigation of the urinary tract to rule out obstructive uropathy or renal duplication anomaly. A transient and momentary dilatation of the ureter is a fortuitous finding that demonstrates the unobstructed passage of urine from kidney to bladder (Fig. 2.5).

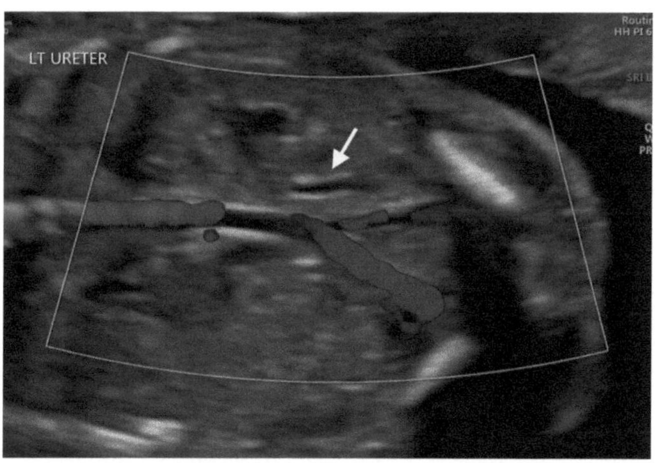

Figure 2.5 Normal ureter *(arrow)*: incidental observation of transitory passage of urine from kidney to bladder on a 20-week fetus.

Bladder

Confirmation of fetal urine production is possible on 2D, grayscale imaging in the first trimester only when the bladder starts to distend. High-resolution ultrasound imaging has permitted the bladder to be seen at 9 weeks on transvaginal imaging and at 11 weeks transabdominally (Abramovich, 1970; Bronshtein, Yoffe, & Brandes, 1990). The fetal bladder can be seen as a fluid-filled spherical space in the fetal pelvis on sagittal and transverse views of the first trimester fetus. Color Doppler imaging reveals two umbilical arteries flanking the bladder bilaterally (Fig. 2.6A). Sebire established that the longitudinal diameter of the fetal bladder between 11 and 14 weeks gestational age should be less than 7 mm (Sebire, Von Kaisenberg, Rubio, Snijders, & Nicolaides, 1996). A measurement more than the 7 mm cut-off in the first trimester would be suggestive of megacystis and could be an early signal of lower urinary tract obstruction, requiring further evaluation in the second trimester (Fig. 2.6B). Longitudinal bladder dimensions greater than 12 mm in the first trimester, may also be a marker of chromosomal abnormalities, particularly, Trisomy 18, 21, or 13 (Liao, Sebire, Geerts, Cicero, & Nicolaides, 2003; Syngelaki, Guerra, Ceccacci, Efeturk, & Nicolaides, 2017). Criteria for normal bladder size on sonography do not appear to vary over gestation and can be ascertained by measuring the sagittal bladder length in millimeter which should equate to the gestational age in weeks plus two (Abramovich, 1970; Maizels et al., 2004).

It should be noted that the presence of amniotic fluid on ultrasound imaging, an indirect marker of fetal urine production, while reassuring, cannot confirm normal renal function in the fetus in the first trimester as it is the placenta, rather than the fetal kidneys that is the major contributor to amniotic fluid production at this stage of pregnancy (Mileto et al., 2018).

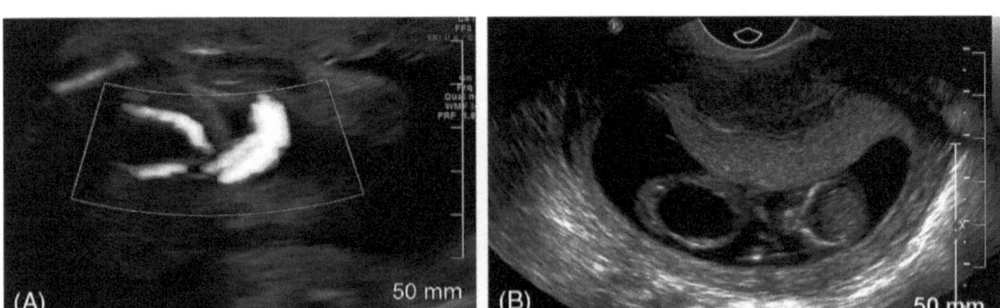

Figure 2.6 (A) Color Doppler representation of umbilical arteries flanking the fetal bladder at 25 weeks. (B) Megacystis in a 13-week fetus diagnosed with LUTO. On coronal view, the longitudinal bladder diameter measures more than 7 mm.

The observation of normal amniotic fluid volume surrounding the fetus in the second and third trimesters is reassuring that at least one kidney is functional as fetal urination contributes an increasing amount to total amniotic fluid volume in the latter two trimesters of pregnancy. The fetal kidneys produce approximately 2–5 mL/h at 22 weeks of gestation and 30–50 mL/h at 40 weeks (Lee et al., 2007). Other references define normal fetal diuresis as 500–700 cc of amniotic fluid per day in the last trimester of pregnancy (van Otterlo, Wladimiroff, & Wallenburg, 1977) and sonographic demonstration of a vertical pocket of fluid measuring at least 2 cm suggests adequate fetal urination is occurring (Mileto et al., 2018).

Prenatal diagnosis of ureterocele

Congenital anomalies affect 3%–5% of children born and can carry with them a lifetime burden of disease. Congenital anomalies of the kidneys and urinary tract (CAKUT) contribute about 20% of the tally, manifesting as nephropathies or uropathies (Mattoo, 2011). Examination of the fetal bladder can give important clues to the presence of abnormalities of either the upper or lower urinary tract.

Sonographic demonstration of a well-delineated thin-walled cystic structure seen within the fetal bladder, at the level of the ureterovesical junction (UVJ) is pathognomonic of ureterocele. It can be most easily appreciated in axial view of the fetal pelvis; however, diagnosis is possible on sagittal and transverse views of the fetus (Fig. 2.7A). Three-dimensional (3D) imaging may be useful adjunct to demonstrate spatial relationship of ureterocele within the bladder (Fig. 2.7B) (Didier, Chow, Kwatra, Retik, & Lebowitz, 2017). It represents the dilatation of the intravesical submucosal ureter, associated with duplication of the ipsilateral ureter up to the renal pelvis in 85%–90% of cases. It may be associated or not with hydronephrosis.

Athey is credited with the first antenatal sonographic demonstration of ureterocele in 1983 in a fetus at 22 weeks' gestational age (Athey, Carpenter, Hadlock, & Hendrick, 1983). They described membranes as "multiseptations" within the fetal bladder, dividing it in three compartments, along with development of hydronephrosis noted on the third trimester. Given the current advances in prenatal imaging, ureteroceles can be easily diagnosed by ultrasound as early as 16 weeks' gestation (Godinho et al., 2013).

Ureteroceles are more frequently seen in female fetuses (4:1), and are frequently associated with other genitourinary anomalies, such as renal duplex collecting system. In males it can be seen as an isolated finding

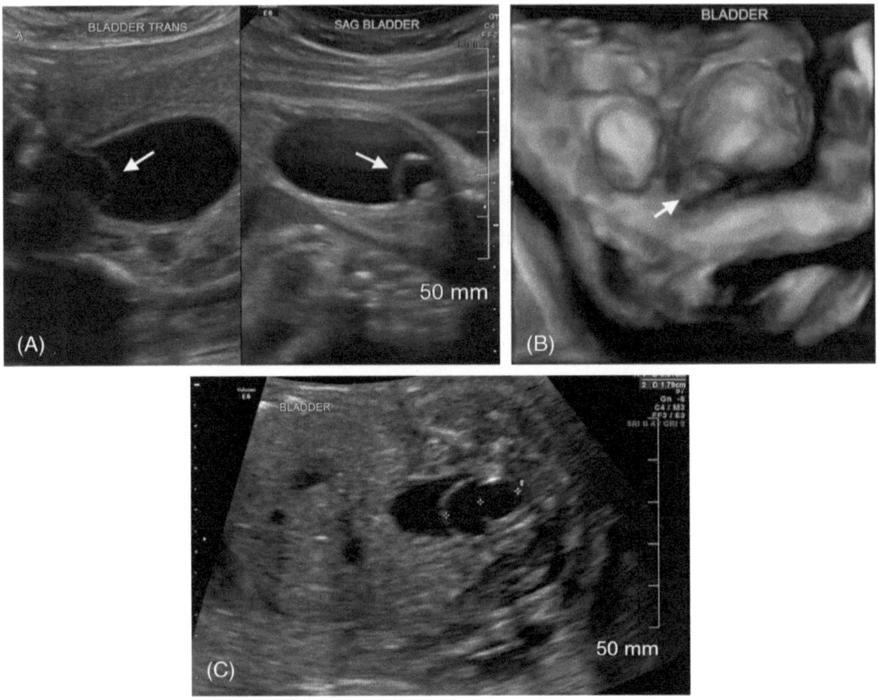

Figure 2.7 (A) Ureterocele in a 29-week fetus, 2D transverse and sagittal views *(arrows)*. (B) 3D image demonstrates saccular structure within the bladder consistent with ureterocele *(arrow)*. (C) Double ureterocele in a 25-week fetus, giving a multiseptated appearance to the bladder.

(Didier, Chow, Kwatra, Retik, & Lebowitz, 2017). The laterality shows a slight predominance for the left side, and nearly 10% of case present as bilateral disease. Approximately 80% of ureteroceles originate from the duplicated upper portion of the ureter that connects to the duplicated upper renal moiety. In these cases, the upper pole tends to be dysplastic or poorly functioning (Schultz & Yoneka Toda, 2016). Ureteroceles can be unilateral or bilateral. In bilateral cases, the bladder can exhibit multiseptated appearance secondary to distension of bilateral ectopic ureters within (Fig. 2.7C) (Adiego et al., 2017).

Based on the Weigert and Meyer rule, (Meyer, 1907; Weigert, 1877), the upper moiety of a duplex collecting system drains via a separate ureter that inserts ectopically and inferiorly to the orthotopic uterovesical junction in the bladder. Outside the bladder, the ectopic ureter can insert anywhere along an "ectopic pathway." In males, the ureter can insert into the bladder, prostatic urethra, or genital ducts. In females, the ectopic ureter can insert

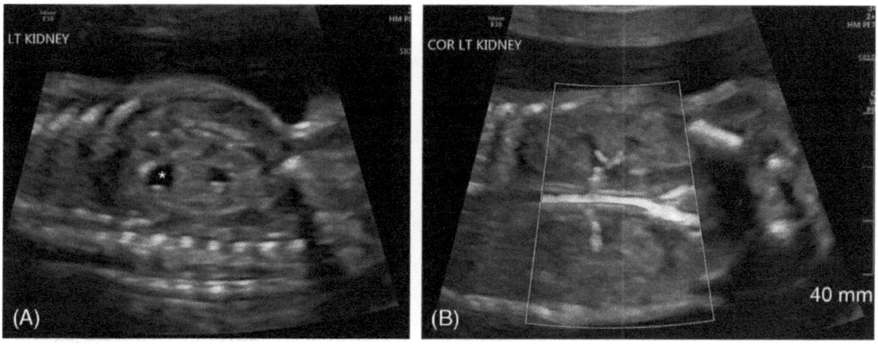

Figure 2.8 Duplex kidney on sagittal view (A) dilated upper renal pelvis (star); (B) branching of the left renal artery supplying upper and lower renal poles on a 19-week fetus.

into the urethra, vagina, uterus, or the fallopian tubes (Didier et al., 2017). The ectopic implantation outside the bladder may be difficult to appreciate on prenatal sonography, however, fetal perineal sonography has been reported to aid in diagnosis of ectopic ureteric implantation into the urethra of both male and female fetuses (Boopathy Vijayaraghavan, 2007).

Duplication of the renal collecting system can be diagnosed in the second trimester, when on sagittal view of the abdomen two distinct and noncommunicating renal pelvicalyceal systems, are noted (Fig. 2.8A). Other sonographic signs of duplex collecting system include bifurcated renal artery with vascularization feeding upper and lower renal poles (Fig. 2.8B), renal length ≥ 95th centile for GA, the presence of a cyst-like structure in the upper pole of the kidney, a dilated ipsilateral ureter, and a ureterocele (Abuhamad, Horton, Horton, & Evans, 1996).

Once duplex collecting system is suspected, sonographic examination of the fetal bladder should attempt to rule out ureterocele and vice versa (Ben Hamouda et al., 2017). Some degree of hydronephrosis can usually be seen when an ureterocele is present (Fig. 2.9A and B). When that happens, dilatation of the upper renal pelvis can be appreciated while the lower pelvis can remain normal. Association with VUR of the *lower* moiety has been described in cases of ureterocele with double collecting system.

One of the pitfalls of imaging ureteroceles is that it is a dynamic structure, and as such, it can be easily missed during the second trimester anatomy ultrasound if the fetal bladder is overfilled. Overdistention of fetal bladder with urine may push the ureterocele against the bladder wall causing urine within its sac to flow back into the ureter, decompressing its walls. As with most fetal anomalies, ureterocele can also be easily missed if there is

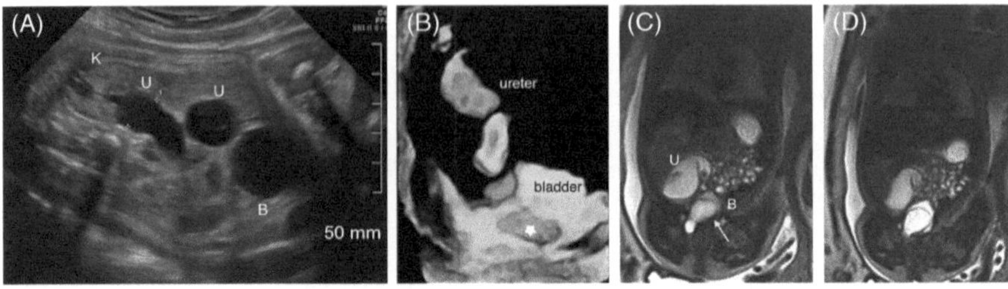

Figure 2.9 (A) 2D and 3D ultrasound views of hydronephrosis in a 29-week fetus. (*K*, kidney; *U*, tortuous hydroureter; *B*, bladder) (B) The ureterocele can be noted as a negative image in the bladder on 3D inversion mode *(star)* (C) and (D) Fetal MRI T2-weighted-sequences: coronal view confirming ureterocele (*U*, tortuous hydroureter; *B*, bladder; *arrow*, large ureterocele demonstrated by large cystic mass within the fetal bladder). *Image (C) courtesy of Dr. Elka Miller (CHEO, Children's Hospital of Eastern Ontario, Ottawa, ON, Canada).*

acoustic shadowing from overlying of bones preventing visualization of the bladder, especially in the third trimester of pregnancy.

Genetics

Urinary tract formation is a complex process, dependent of the interaction of multiple genes. Unlike other CAKUT, ureterocele has no known genetic association to date. Since 2005, 10 families have been described with multiple cases of ureterocele in both single and double systems, suggesting the possibility of genetic predisposition. Development of ureterocele can be a consequence of aberrations in multiple processes during embryological development from the initial morphogenesis of the ureter, and including its migration and incorporation to the cloacal plate, resulting from interaction of numerous morphological, cellular, and biochemical factors (Schultz & Yoneka Toda, 2016).

Differential diagnosis

The finding of a cystic mass in the fetal bladder in prenatal sonography is a unique diagnostic feature of ureterocele. Upon recognition of this entity, a concentrated anatomic survey of the fetal urinary tract should be undertaken to differentiate its various etiologies, which can indicate pathologies of the upper or lower urinary tract. Rare cases have been described of a very large ureterocele that had its walls prolapsed into the urethra, causing obstruction and consequently megacystis and oligohydramnios (Austin, Cain, Casale, Hiett, & Rink, 1998).

Other diagnostic modalities

Ultrasound is an excellent first-line imaging tool to diagnose ureterocele, providing accessible and cost-effective prenatal diagnosis. Through its ability to provide early and differential diagnosis and rule out associated anomalies, it supplies valuable information to the clinician to inform parental counseling and plan appropriate monitoring throughout the pregnancy. When ultrasound is equivocal or more specific information is needed, other methods can be proposed to clarify diagnosis.

Fetal MRI is an excellent diagnostic adjunct because of its ability to provide high-quality tissue contrast and wider anatomic coverage, especially in complex fetal anomalies. The reported disadvantages of ultrasound imaging, namely its dependence on the user and other imaging limitations such as volume of amniotic fluid, fetal position, and maternal body habits, are typically not characteristic of MRI. Its high specificity in assessing fetal anatomy is of great value when assessing for associated anomalies. It is a reliable diagnostic tool and aids the perinatology team when planning management and counseling parents (Martllln et al., 2004).

The normal fetal renal parenchyma is of intermediate signal intensity on T2-weighted images, whereas the renal collecting systems are of higher signal intensity. The fetal urinary bladder is a high-signal-intensity round or ovoid structure in the pelvis. If there is hydronephrosis due to ureterocele or ureteral ectopic voiding, the dilatation of the ureter will be depicted as a high- signal tubular structure on T2-weighted sequences and as a low-signal structure on T1-weighted images. The ureterocele is easily identified as a round structure inside the bladder (Fig. 9C and D). However, fetal MRI is not capable of identifying precisely the condition and extension of ureterocele into the urethra, nor does it offer visualization of the bladder surface. A new MRI technique has been described: fetal MRI virtual cystoscopy (Abdol-Mohammad Kajbafzadeh, Nabavizadeh, Beigi, Alinia, & Mirshahvalad, 2018). It is a 3D fetal MRI that provides detailed multi-slice images of the bladder and ureters, in its whole extension. Three-dimensional virtual cystoscopy provides a non-invasive evaluation of fetal lower urinary tract obstruction that in the future may help in the selection of candidates for fetal cystoscopy (Werner, Lopes, Ribeiro, 2017). More studies are needed to establish virtual cystoscopy role on the antenatal management of ureteroceles.

Prenatal diagnosis of ureteroceles is essential to plan early multidisciplinary care to avoid long-term renal consequences. In essence, it is of crucial importance to determinate the distal extension of the ureterocele and the possibility of lower urinary tract obstruction (LUTO) in order to plan both fetal and

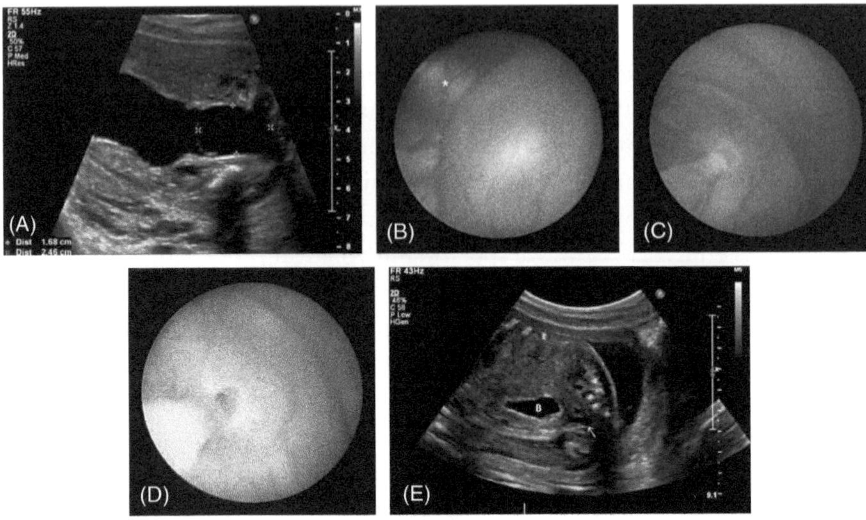

Figure 2.10 (A) Ureterocele on 2D axial view of fetal bladder. (B) Fetal cystoscopy view of dilated ureter and ureterocele *(star)*. (C) Laser on ureterocele. (D) Laser puncture fenestration of ureterocele by Colander technique. (E) Post-procedure ultrasound of fetal bladder, ureterocele resolved. *Images courtesy of Dr. Greg Ryan (Ontario Fetal Centre, Mount Sinai Hospital, Toronto, ON, Canada).*

neonatal management (Upadhyay et al., 2002). Early ultrasound diagnosis can direct appropriate and timely referral to pediatric urologists and fetal medicine specialists for consideration of fetal therapy in select cases. Importantly, it facilitates optimal postnatal planning, including adequate institution of antibiotic prophylaxis, to reduce short and long-term morbidity associated with recurrent urinary tract infections and to triage patients who are candidate for enhanced renal surveillance in the newborn period (Fig. 2.10).

References

Abdol-Mohammad Kajbafzadeh, Nabavizadeh, B., Beigi, R. S. H., Alinia, P., & Mirshahvalad, S. A. (2018). Virtual three-dimensional magnetic resonance fetal cystoscopy: a novel modality for precise in utero evaluation of urinary tract. *Urology Journal* https://doi.org/10.22037/uj.v0i0.4569.

Abramovich, D. R. (1970). Fetal factors influencing the volume and composition of liquor amnii. *The Journal of Obstetrics and Gynaecology of the British Commonwealth*, 77(10), 865–877.

Abuhamad, A. Z., Horton, C. E., Horton, S. H., & Evans, A. T. (1996). Renal duplication anomalies in the fetus: clues for prenatal diagnosis. *Ultrasound in Obstetrics and Gynecology*, 7, 174–177.

Adiego, B., Martinez-Ten, P., Perez-Pedregosa, J., Illescas, T., Barron, E., Wong, A. E., & Sepulveda, W. (2017). Antenatally diagnosed renal duplex anomalies. *Journal of Ultrasound in Medicine*, *30*(6), 809–815 https://doi.org/10.7863/jum.2011.30.6.809.

AIUM–ACR–ACOG–SMFM–SRU Practice Parameter for the Performance of Standard Diagnostic Obstetric Ultrasound Examinations.(2018). *J Ultrasound Med*, *37*, E13-E24. https://doi.org/10.1002/jum.14831.

Athey, P., Carpenter, R., Hadlock, F., & Hendrick, T. (1983). Ultrasonic demonstration of ectopic ureterocele. *Pediatrics*, *71*(4), 568–571.

Austin, P. F., Cain, M. P., Casale, A. J., Hiett, A. K., & Rink, R. C. (1998). Prenatal bladder outlet obstruction secondary to ureterocele. *Urology*, *52*(6), 1132–1135 https://doi.org/10.1016/S0090-4295(98)00337-9.

Ben Hamouda, H., Bouchahda, H., Ghanmi, S., Wannes, S., Soua, H., Hamza, H., & … Sfar, M. T. (2017). Diagnostic prénatal et prise en charge de deux cas d'urétérocèles bilatéraux sur uretères simplex. *Archives de Pediatrie*, *24*(9), 860–864 https://doiorg/10.1016/jarcped.2017.06.001.

Boopathy Vijayaraghavan, S. (2007). Fetal perineal sonography in the diagnosis of ectopic ureteric opening. *Ultrasound in Obstetrics and Gynecology*, *29*(3), 347–351 https://doi.org/10.1002/uog.3937.

Bronshtein, M., Yoffe, N., & Brandes, J. (1990). First and early second-trimester diagnosis of fetal urinary tract anomalies using transvaginal sonography. *Prenatal Diagnosis*, *10*, 653.

Chalmers, D. J., Meyers, M. L., Brodie, K. E., Palmer, C., & Campbell, J. B. (2016). Inter-rater reliability of the APD, SFU and UTD grading systems in fetal sonography and MRI. *Journal of Pediatric Urology*, *12*(5), 305.e1-305.e5. https://doi.org/10.1016/j.jpurol.2016.06.012.

Chapman, T. (2012). Fetal genitourinary imaging. *Pediatric Radiology*, *42*(SUPPL. 1), 115–123 https://doiorg/10.1007/s00247-011-2172-6.

Darahem Tedesco, G., de Silva Bussamra, L., Araujo Júnior, E., Schwach Werneck Britto, I., Marcondes Machado Nardozza, L., Fernandes Moron, A., & Aoki, T. (2009). Reference range of fetal renal volume by three-dimensional ultrasonography using the vocal method. *Fetal Diagnosis and Therapy*, *25*, 385–391.

Devriendt, A., Cassart, M., Massez, A., Donner, C., & Avni, F. E. (2013). Fetal kidneys: additional sonographic criteria of normal development. *Prenatal Diagnosis*, *33*, 1248–1252.

Didier, R. A., Chow, J. S., Kwatra, N. S., Retik, A. B., & Lebowitz, R. L. (2017). The duplicated collecting system of the urinary tract: embryology, imaging appearances and clinical considerations. *Pediatric Radiology*, *47*(11), 1526–1538 https://doi.org/10.1007/s00247-017-3904-z.

Fernbach, S., Maizels, M., & Conway, J. (1993). Ultrasound grading of hydronephrosis: introduction to the system used by the Society for Fetal Urology. *Pediatric Radiology*, *23*(6), 478–480.

Godinho, A. B., Nunes, C., Janeiro, M., Carvalho, R., Melo, M. A., & Da Graça, L. M. (2013). Ureterocele: antenatal diagnosis and management. *Fetal Diagnosis and Therapy*, *34*(3), 188–191 https://doi.org/10.1159/000353388.

Gramellini, D., Fieni, S., Caforio, E., Benassi, G., Bedocchi, L., & Beseghi, U. (2006). Diagnostic accuracy of fetal renal pelvis anteroposterior diameter as a predictor of significant postnatal nephrouropathy: Second versus third trimester of pregnancy. *American Journal of Obstetrics and Gynecology*, *194*, 167–173.

Hershkovitz, R., Amichay, K., & Stein, G. Y. (2011). The echogenicity of the normal fetal kidneys during different stages of pregnancy determined objectively. *Archives of Gynecology and Obstetrics*, 807–811 284.

International Society of Ultrasound in Obstetrics and Gynecology (ISUOG) practice guidelines: Performance of first-trimester fetal ultrasound scan. (2013). In: *Ultrasound in Obstetrics and Gynecology* (Vol. 41, pp. 102–113). https://doi.org/10.1002/uog.12342.

Lee, S. M., Park, S. K., Shim, S. S., Jun, J. K., Park, J. S., & Syn, H. C. (2007). Measurement of fetal urine production by three-dimensional ultrasonography in normal pregnancy. *Ultrasound in Obstetrics and Gynecology*, *30*(3), 281–286.

Liao, A. W., Sebire, N. J., Geerts, L., Cicero, S., & Nicolaides, K. H. (2003). Megacystis at 10-14 weeks of gestation: chromosomal defects and outcome according to bladder length. *Ultrasound in Obstetrics and Gynecology*, *21*, 338–341.
Maizels, M., Alpert, S., Houston, J., Sabbagha, R., Parilla, B., & MacGregor, S. (2004). Fetal bladder sagittal length: a simple monitor to assess normal and enlarged fetal bladder size, and forecast clinical outcome. *Journal of Urology*, *172*(5), 1995–1999.
Martllln, C., Darnell, A., Durllln, C., Bermllldez, P., Mellado, F., & Rigol, S. (2004). Magnetic resonance imaging of the intrauterine fetal genitourinary tract: normal anatomy and pathology. *Abdominal Imaging*, *29*(3), 286–302 https://doi.org/10.1007/s00261-003-0105-z.
Mattoo, T. K. (2011). Diagnosis of congenital renal anomalies in children. *Clinical Biochemistry*, *44*(7), 498 https://doi.org/10.1016/j.clinbiochem.2011.03.131.
Meyer, R. (1907). Zur Anatomie und entwicklungsgeschichte der ureterverdoppelung. [The anatomy and developmental history of the duplicated ureter.]. *Virchows Archiv Für Pathologische. Anatomie Und Physiologie Und Für Klinische Medizin*, *87*, 408.
Mileto, A., Itani, M., Katz, D., Siebert, J., Dighe, M., Dubinsky, T., & Moshiri, M. (2018). Fetal urinary tract anomalies. review of pathophysiology, imaging, and management. *The American Journal of Roentgenology, Radium Therapy, and Nuclear Medicine*, *210*(5), 1010–1021.
Nguyen, H., Benson, C., Bromley, B., Campbell, J., Chow, J., Coleman, B., & Crino, J. (2014). Multidisciplinary consensus on the classification of prenatal and postnatal urinary tract dilation (UTD classification system). *Journal of Pediatric Urology*, *10*(6), 982–998.
Schultz, K., & Yoneka Toda, L. (2016). Genetic basis of ureterocele. *Current Genomics*, *17*(1), 62–69 https://doiorg/10.2174/1389202916666151014222815.
Sebire, N. J., Von Kaisenberg, C., Rubio, C., Snijders, R. J., & Nicolaides, K. H. (1996). Fetal megacystis at 10-14 weeks of gestation. *Ultrasound in Obstetrics and Gynecology*, *8*, 387–390 https://doi.org/10.1046/j.1469-0705.1997.08060387.x.
Shin, J., Seo, Y., Kim, J., & Park, K. (2007). Nomogram of fetal renal growth expressed in length and parenchymal area derived from ultrasound images. *Journal of Urology*, *178*(5), 2150–2154.
Syngelaki, A., Guerra, L., Ceccacci, I., Efeturk, T., & Nicolaides, K. H. (2017). Impact of holoprosencephaly, exomphalos, megacystis and increased nuchal translucency on first-trimester screening for chromosomal abnormalities. *Ultrasound in Obstetrics and Gynecology*, *50*, 45–48.
Upadhyay, J., Bolduc, S., Braga, L., Farhat, W., Bägli, D. J., McLorie, G. A., & El-Ghoneimi, A. (2002). Impact of prenatal diagnosis on the morbidity associated with ureterocele management. *Journal of Urology*, *167*(6), 2560–2565 https://doi.org/10.1016/S0022-5347(05)65037-9.
van Otterlo, L., Wladimiroff, J., & Wallenburg, H. (1977). Relationship between fetal urine production and amniotic fluid volume in normal pregnancy and pregnancy complicated by diabetes. *British Journal of Obstetrics and Gynaecology*, *84*(3), 205–209.
van Vuuren, S. H., Damen-Elias, H. A., Stigter, R. H., van der Doef, R., Goldschmeding, R., de Jong, T. P., & Pistorius, L. R. (2012). Size and volume charts of fetal kidney, renal pelvis and adrenal gland. *Ultrasound in Obstetrics and Gynecology*, *40*, 659–664.
Weigert, C. (1877). Über einige Bildungsfehler der Ureteren (Some malformations of the ureters). *Virchows Archiv Für Pathologische Anatomie Und Physiologie Und Für Klinische Medizin*, *70*, 490.
Werner, H., Lopes, J., Ribeiro, G., et al. (2017). Three-dimensional virtual cystoscopy: non-invasive approach for the assessment of urinary tract in fetuses with lower urinary tract obstruction. *Prenatal Diagnosis*, *37*, 1350–1352.
Yu, C. -H., Chang, C. -H., Chang, F. -M., Ko, H. -C., & Chen, H. -Y. (2000). Fetal renal volume in normal gestation: a three-dimensional ultrasound study. *Ultrasound in Medicine and Biology*, *26*(8), 1253–1256.

CHAPTER 3

Prenatal management of ureteroceles

Julia Beth Finkelstein[a], **Jason P. Van Batavia**[b]
[a]Pediatric Urology, Department of Urology, Boston Children's Hospital, Boston, MA, United States
[b]Pediatric Urology, The Children's Hospital of Philadelphia, Division of Urology, Perelman School of Medicine, University of Pennsylvania, Philadelphia, PA, United States

Background/Introduction

Some ureteroceles are small and of little clinical significance. Other ureteroceles are large, presenting as balloon-like lesions that can cause bladder outlet obstruction or protrude to the contralateral side, causing bilateral ureterovesical obstruction (Austin et al., 1998; Hansen, Cooper, & Yankowitz, 2002). As long as the amniotic fluid level remains normal and fetal lung development is not compromised, the prenatal management of ureteroceles is generally conservative and the ureterocele is monitored and/or treated postnatally (Ashmead et al., 2004a; Quintero, Homsy, Bornick, Allen, & Johnson, 2001). Shankar and colleagues demonstrated that, when managed expectantly, 27% of children with an antenatal diagnosis of ureterocele do not seem to have any issues after birth (Shankar, Vishwanath, & Rickwood, 2001).

However, in rare cases, a large ureterocele can result in bladder outlet obstruction with subsequent oligo- or anhydramnios, which can cause bilateral renal damage, pulmonary hypoplasia, and potentially early neonatal death (Ashmead et al., 2004a; Quintero et al., 2001). In these circumstances, the goal of fetal intervention is to prevent further renal damage from obstruction but more importantly to increase amniotic fluid levels to facilitate pulmonary development and function. The link between urine production and lung development rests on the fact that after the first trimester of gestation, >90% of the amniotic fluid volume is composed of urine. Thus blockage of urine production from a large ureterocele that drops into the bladder neck or proximal urethra can lead to oligo- or anhydramnios.

Different approaches to prenatal treatment of ureteroceles have been described in mostly single institution case reports. Given the limited evidence in the literature and the highly specialized nature of the interventions, there

Ureteroceles
http://dx.doi.org/10.1016/B978-0-12-817888-1.00003-3

is currently no standardized guideline on when and how to intervene on the prenatally diagnosed ureterocele. When fetal intervention is attempted, first-line treatment generally involves minimally invasive techniques, primarily incision of the ureterocele via laser or needle puncture. Though more commonly performed at specialized fetal centers for bladder outlet obstruction secondary to posterior urethral valve, vesicoamniotic shunting is generally reserved for when there is failure of decompression of the ureterocele after other attempts at fetal intervention. It must be noted that all of these intrauterine procedures are associated with the risk of preterm rupture of membranes (Beck, Lewi, Gucciardo, & Devlieger, 2012).

Since the experience with and outcomes of fetal intervention is limited, the benefits are unclear. Indeed, there is no definitive evidence that early decompression of fetal hydroureteronephrosis restores renal function (Farrugia, Braun, Peters, Ruano, & Herndon, 2017). As Smith-Harrison and colleagues note, in utero intervention may improve perinatal survival but may not alter the long-term prognosis (Smith-Harrison, Hougen, Timberlake, & Corbett, 2015).

Prenatal surveillance

When ureteroceles cause bladder outlet obstruction, serial ultrasound scans are necessary. Amniotic fluid volume, bladder volume, the degree of hydronephrosis, and echogenicity of the kidneys must be monitored (Soothill, Bartha, & Tizard, 2003). In addition, some investigators have proposed the use of one or more fetal urinary biochemistry markers as a surrogate for fetal renal function including sodium, chloride, calcium, osmolarity, and total protein (Table 3.1) (Adzick & Harrison, 1994; Glick et al., 1985; Ruano et al., 2016). From their review of 23 studies, Morris and colleagues

Table 3.1 Fetal urine chemistry values thought to be associated with poor fetal renal function (Adzick & Harrison, 1994; Glick et al., 1985; Ruano et al., 2016).

	Fetal urine value suggestive of poor renal function
Sodium	>100 mg/dL
Chloride	>90 mg/dL
Calcium	>100 mg/dL
Osmolarity	>200 mg/dL
Total protein	>20 mg/dL
β2-microglobulin	>4.0 mg/dL

noted that fetal urinary calcium and sodium were the two most accurate biochemistry tests of fetal renal function (Morris, Quinlan-Jones, Kilby, & Khan, 2007).

In addition, the use of fetal urinary ß2-microglobulin has been suggested as a marker of renal tubular damage since this microglobulin has been noted to be excreted in the urine of patients with postnatal renal disease (Tassis et al., 1996). Despite some early supportive studies, a subsequent systematic review in 2007 determined that there was insufficient evidence to support the use of fetal urinary ß2-microglobulin as a predictor of fetal renal function (Morris et al., 2007). Interestingly, Spaggiari and associates analyzed ß2-microglobulin levels in fetal serum in 42 fetuses with bilateral or lower urinary tract obstruction. The authors found that sequential measurements of serum fetal ß2-microglobulin predicted postnatal renal outcome more accurately than a single test (Spaggiari et al., 2017). Another, more recent, study by Spaggiari and colleagues compared fetal serum and fetal urine biochemistry markers in predicting postnatal renal outcomes in fetuses with lower urinary tract obstruction. In this study, the authors noted that both fetal serum and urine markers, particularly ß2-microglobulin levels, had similar prognostic values in predicting renal outcomes (Spaggiari et al., 2018). While none of these studies were focused on prenatal ureteroceles, but rather included all diagnoses of lower urinary tract obstruction (LUTO), the limited number of studies on prenatal evaluation for any diagnosis of LUTO requires generalization across all diagnoses at this time.

Besides, ideal timing for first and repeat fetal urinary biochemistry sampling is largely unknown, but may be based on the known embryology of the fetal kidney in which metanephric development begins between weeks 4–5 of gestation and nephron development starts around 10–12 weeks. Since amniotic fluid after the first trimester is composed primarily of fetal urine, sampling starting in the second trimester around 18 weeks of gestation, if indicated, seems reasonable to reflect fetal renal function. Serial fetal urine aspiration for biochemistry levels may provide more value than a single time point but frequency of repeat aspiration requires future study (Johnson et al., 1995).

In 2016, the Texas Children's Fetal Center developed a lower urinary tract obstruction (LUTO) classification system to improve the delineation of patients who may benefit from fetal intervention (Farrugia et al., 2017; Ruano et al., 2016). This system includes: fetal urinary biomarkers suggestive of renal injury, amniotic fluid levels, and imaging studies to identify renal dysplastic changes (Table 3.2). According to their staging classification, fetuses with low

Table 3.2 Texas Children's Fetal Center standardized multidisciplinary prenatal staging and fetal intervention recommendations (Ruano et al., 2016).

	Stage 1 (mild LUTO)	Stage 2 (severe LUTO with findings suggestive of preserved fetal renal function)	Stage 3 (severe LUTO with findings suggestive of abnormal fetal renal function)
Amount of amniotic fluid	Normal	Oligo- or anhydramnios	Oligohydramnios, but usually anhydramnios
Echogenicity of fetal kidneys	Normal	Hyperechogenic	Hyperechogenic
Renal cortical cysts	Absent	Absent	Can be present
Renal dysplasia	Absent	Absent	Can be present
Fetal urinary biochemistry	Favorable	Favorable within three consecutive evaluations	Not favorable after three consecutive evaluations
Fetal intervention	Not indicated	Indicated to prevent pulmonary hypoplasia and severe renal impairment	May be indicated to prevent pulmonary hypoplasia but not postnatal renal impairment

risk (stage 1) are not recommended for intervention but instead undergo surveillance. For fetuses with signs of severe LUTO but without evidence of severe pre-existing renal injury (stage 2), intervention is recommended as these patients are thought to be at high risk for development of renal damage and/or pulmonary hypoplasia. For the most ill fetuses with severe LUTO and findings suggestive of severe fetal renal damage, intervention may be recommended to prevent pulmonary hypoplasia with the understanding that intervention is unlikely to improve postnatal renal function (Farrugia et al., 2017).

Fetuses with normal amniotic fluid and particularly those with stage 1 LUTO as per the Texas Children's classification may safely undergo surveillance during pregnancy. Repeat fetal ultrasounds to monitor amniotic fluid levels, renal parenchymal characteristics, and urinary tract dilation, should be performed throughout pregnancy although specific timing may vary based on institution and individual fetal findings. In cases where there is concern for abnormal fetal renal function, fetal urine biochemistry markers may be obtained and help decide on continued surveillance versus potential intervention.

Prenatal intervention

Timing of intervention

Unfortunately, the ideal timing for fetal intervention to preserve renal and/or pulmonary function is unknown. Several studies have shown that the timing of oligo- or anhydramnios is associated with the ultimate outcome (Mandell, Peters, Estroff, & Benacerraf, 1992; Peters et al., 1991). Development of late oligohydramnios after 30 weeks of gestation secondary to a urological issue, will not likely impact pulmonary function or cause pulmonary hypoplasia (Mandell et al., 1992; Shipp, Bromley, Pauker, Frigoletto, & Benacerraf, 1996). Conversely, oligohydramnios before 24 weeks of gestation has the potential for severe pulmonary hypoplasia and subsequent postnatal respiratory distress and/or death. As mentioned earlier, fetal renal development begins early in the first trimester and nephron development commences around the 10th week of gestation. Therefore, any fetal intervention to preserve renal function must also likely occur early in gestation, with some authors suggesting before 22–23 weeks of gestation (Farrugia et al., 2017). Past this critical point, whenever it may be, renal damage is likely irreversible and thus intervention should only be undertaken to benefit pulmonary development.

Taken all together, an embryological understanding of renal and pulmonary development suggests that fetal intervention, when warranted, will likely provide the most benefit during the second trimester. In their proposal for standardized prenatal management based on disease severity, Ruano and colleagues from Texas Children's Fetal Center offered fetal intervention between 16 and 34 weeks gestation (Farrugia et al., 2017). Whether a more restrictive cutoff (e.g., 30 weeks) is safer based on benefits/risks is yet to be determined and may be answered in future studies.

Types of intervention

Vesicoamniotic shunt

Fetal lower urinary tract obstruction has been relieved by bladder drainage via placement of a vesicoamniotic shunt that allows for continuous drainage into the amniotic cavity. These shunts can easily become displaced or obstructed. In addition, the procedure can be associated with premature labor, rupture of membranes, infection and bleeding (Soothill et al., 2003). Still, a meta-analysis of vesicoamniotic shunting for LUTO did favor intervention (Nassr et al., 2017).

In a randomized multi-center trial, PLUTO, the fetal bladder was visualized with an ultrasound and, with a sterile, "minimum-touch technique," a vesicoamniotic pigtail catheter was inserted percutaneously (Morris et al., 2013). Placement of the shunt was confirmed immediately and at several hours after the procedure. Though the indication was not limited to obstruction from ureterocele, the PLUTO study results supported improved perinatal survival in the intervention arm (Morris et al., 2013). However, the investigators noted that "the size and direction of the effect remained uncertain, such that benefit could not be conclusively proven." Furthermore, there was no difference in the incidence of end-stage renal disease (Morris et al., 2013).

Percutaneous decompression

In their case report, Hansen and associates describe the successful treatment of a large intravesical ureterocele with single percutaneous needle decompression (Hansen et al., 2002). Continuous ultrasound guidance was used to place a 22-gauge needle percutaneously into the bladder and ureterocele. This group reports performing a single puncture with subsequent manipulation of the needle being necessary to drain both the bladder and ureterocele. After the intervention, the ureterocele appeared decompressed and, 3 days later, the amniotic fluid level was normal. The ureterocele continued to appear decompressed on ultrasounds throughout the duration of the pregnancy (Hansen et al., 2002).

Similarly, Crane reported on the successful decompression of a ureterocele by in utero needle puncture under ultrasound guidance at 25 weeks' gestation for severe oligohydramnios (Crane, 2000). All repeat ultrasounds were normal for the remainder of the pregnancy and delivery was induced at 39 weeks' gestation. The neonate was noted to void spontaneously at birth and had normal urinary electrolytes at 3 hours of age. Given postnatal ultrasound findings of prolapsed ureterocele and bilateral hydroureteronephrosis, the neonate underwent cystoscopy and ureterocele unroofing on day of life 1. Repeat renal ultrasound at 6 weeks of life, the last imaging described in the case report, showed improvement (Crane, 2000).

Finally, Chalouhi et al report on 10 cases of ureterocele treated prenatally at four different centers. In two of these cases, ultrasound guided puncture of the fetal ureterocele was carried out, before resorting to fetoscopy with laser incision (Chalouhi et al., 2017).

Percutaneous "Colander" therapy

In a case of rapid onset of anhydramnios at 28 weeks' gestation, Ashmead and colleagues describe decompression of a ureterocele using in utero vesicocentesis and multiple "colander" punctures by insertion of a single 20-gauge needle under ultrasound guidance. Approximately, 10 small perforations were made in the dome of the ureterocele, which the authors note to be similar to the holes in a colander (Ashmead et al., 2004b) (representative images in Fig. 10). This management was chosen "to avoid creation of an obstructing flap valve and to minimize the risk of postprocedural reflux." Less than 3 days after the intervention, the amniotic fluid level was noted to be in the normal range. However, at 37 weeks' gestation, the patient was found to have recurrent oligohydramnios and the ureterocele was again visualized on ultrasound. Delivery was induced and postnatal ultrasound demonstrated a decompressed ureterocele. A diuretic renogram showed no evidence of obstruction and no immediate surgical treatment was initiated (Ashmead et al., 2004b).

Laser incision of ureterocele

Soothill et al. describe ultrasound-guided laser treatment of a right ureterocele associated with bilateral hydronephrosis and decreased amniotic fluid level in a 16-year-old woman at 29 weeks' gestation (Soothill et al., 2003). The fetal bladder was accessed under ultrasound guidance using a 19-gauge needle and the needle was oriented toward the ureterocele. A 400-μm laser fiber was then introduced through the needle and visualized via ultrasound. Several small incisions were made in the part of the ureterocele farthest away from the normal bladder. Mild urinary ascites was visualized post-procedure. This resolved after 3 days and both hydronephrosis and amniotic fluid index improved. Labor was induced at 38 weeks and postnatal ultrasound showed a normal bladder and mild dilation of bilateral tortuous ureters (Soothill et al., 2003). These authors propose that their method is less invasive than fetoscopy, potentially diminishing fetal as well as maternal morbidity.

Fetoscopy

Fetal cystoscopy was originally described in 1995 for treatment of fetuses with posterior urethral valves (Quintero et al., 1995). Since this report, fetoscopy has more recently been used for laser incision of ureteroceles, as described later. Fetal cystoscopic visualization may permit more precise selection of the size and site of ureterocele puncture. Importantly, it also should protect the bladder wall from inadvertent damage. While considered to be minimally invasive, this procedure can still be associated with rupture of membranes. In addition, the intervention can create urinary fistulas, such as urethro-rectal (Farrugia et al., 2017; Sananes et al., 2015).

Quintero and colleagues report on using fetal cystoscopy to manage a fetal ureterocele complicated by oligohydramnios near 22 weeks' gestation (Quintero et al., 1995). Under ultrasound guidance, the fetal bladder was accessed percutaneously using a 3.5 mm trocar. The ureterocele was incised via fetal cystoscopy and a neodymium:yttrium-aluminum-garnet laser using a 400 μm fiber. This resulted in immediate decompression of the ureterocele and restored the amniotic fluid to normal levels. Postnatal diuretic renography demonstrated a non-functioning right kidney and VCUG revealed grade 5 reflux into a duplex right kidney with megaureters. The infant underwent right nephrectomy and did well post-operatively (Quintero et al., 1995).

Similarly, Montebruno and partners introduced a 3-mm trocar percutaneously into the amniotic cavity and confirmed the presence of prolapsed ureterocele under direct vision with a fetoscope (Montebruno et al., 2015). With combined ultrasound and endoscopic guidance, a diode laser was then used to decompress the ureterocele. Ultrasound showed improvement of bilateral hydronephrosis and the ureterocele was no longer visualized. Postnatal cystography showed bilateral reflux. Because of recurrent urinary tract infections, a right heminephrectomy was performed at 1-year-old and follow up since has been unremarkable (Montebruno et al., 2015).

Persico and associates report on an additional two cases of fetal ureterocele managed with fetal cystoscopic laser decompression (Persico et al., 2017). In the first case, under ultrasound guidance, a 3.3 mm cannula was placed transabdominally into the amniotic cavity and fetal bladder. The right ureterocele was incised endoscopically using a 400-μm diode laser. Follow-up fetal ultrasounds demonstrated substantial improvement in hydronephrosis and normal amniotic fluid volume. Postnatal VCUG revealed high-grade right reflux in the lower system. In the second case, a 1.6-mm "all-seeing needle" was directly introduced under ultrasound guidance and a large ureterocele was visualized. The ureterocele was incised with a 200-μm holmium-YAG laser fiber. Again, follow-up ultrasounds showed improved hydronephrosis and normal amniotic fluid volume. Postnatal VCUG showed bilateral reflux in the lower moieties and no residual evidence of the ureterocele (Persico et al., 2017).

Godinho and partners also describe the case of performing fetoscopy at 30 weeks' gestation due to the progression of bilateral severe hydronephrosis (Godinho et al., 2013). The ureterocele was visualized protruding through the urethra and laser incision of the ureterocele was completed, with partial improvement of hydronephrosis. In this case, the amniotic fluid level was always normal. In postnatal follow up, cystography demonstrated vesicoureteral reflux. After recurrent urinary tract infections, a right heminephrectomy was performed at 1-year-old and the follow up was then unremarkable (Godinho et al., 2013).

Chalouhi et al. describe direct entry into the fetal abdomen and bladder by using the Seldinger technique with a 9Fr trocar in six cases (Chalouhi et al., 2017). A 1.2-mm fetoscope and a 200-um diode laser fiber was then used to incise the ureterocele. Alternatively, in one case, a laser fiber was inserted via a 20-gauge needle under ultrasound guidance and the ureterocele was then incised using the diode laser. In all cases amniotic fluid levels were restored to normal and deliveries resulted in liveborn neonates at a mean gestational age of 38.6 weeks. Serum creatinine levels were also noted to be normal in the first week of life and prenatal intervention was found to provide "complete treatment" of severely obstructing ureteroceles in 80% of cases (Chalouhi et al., 2017).

While these results are encouraging, the lack of clear guideline on evaluation and standardization of treatment options, mixed with the rarity of cases and highly specialized nature of any intervention means that any patient with potential benefit from intervention should be referred in a timely manner to a fetal center with the necessary expertise. This is just as true for severely obstructing ureteroceles as for any case of severe LUTO.

Conclusions

In general, when fetal ureteroceles are diagnosed, the fetus can be managed conservatively with in utero monitoring. Though uncommon, a large ureterocele can bulge or prolapse, causing bladder outlet obstruction. In these rare instances, different management options have been described in mostly single institution case reports. Besides the small sample size, these reports have variable indications for intervention and offer limited follow-up, making it challenging to draw a conclusion about prenatal interventions. The potential fetal benefit must balance the risk associated with intervention. Nevertheless, referral to appropriate tertiary care fetal centers for consultation and potential intervention should be considered in any case of severely obstructing ureterocele with concern for renal and/or pulmonary compromise.

References

Adzick, N. S., & Harrison, M. R. (1994). Fetal surgical therapy. *The Lancet, 343*(8902), 897–902.

Ashmead, G. G., Mercer, B., Herbst, M., Moodley, J., Bota, A., & Elder, J. S. (2004a). Fetal bladder outlet obstruction due to ureterocele. *Journal of Ultrasound in Medicine, 23*, 565–568 https://doi.org/10.7863/jum.2004.23.4.565.

Ashmead, G. G., Mercer, B., Herbst, M., Moodley, J., Bota, A., & Elder, J. S. (2004). Fetal bladder outlet obstruction due to ureterocele: in utero "colander" therapy. *Journal of Ultrasound in Medicine: Official Journal of the American Institute of Ultrasound in Medicine, 23*(4), 565–568.

Austin, P. F., Cain, M. P., Casale, A. J., Kinney Hiett, A., & Rink, R. C. (1998). Prenatal bladder outlet obstruction secondary to ureterocele. *Urology, 52*, 1132–1135 https://doi.org/10.1016/s0090-4295(98)00337-9.

Beck, V., Lewi, P., Gucciardo, L., & Devlieger, R. (2012). Preterm prelabor rupture of membranes and fetal survival after minimally invasive fetal surgery: a systematic review of the literature. *Fetal Diagnosis and Therapy, 31*(1), 1–9.

Chalouhi, G. E., Morency, A. -M., De Vlieger, R., Martinez, J. M., Blanc, T., Hodges, R., ..., & Ville, Y. (2017). Prenatal incision of ureterocele causing bladder outlet obstruction: a multicenter case series. *Prenatal Diagnosis, 37*(10), 968–974.

Crane, J. M. (2000). Fetal bladder outlet obstruction caused by ureterocele. *Obstetrics and Gynecology, 95*(6 Pt 2), 1040.

Farrugia, M. K., Braun, M. C., Peters, C. A., Ruano, R., & Herndon, C. D. (2017). Report on The Society for Fetal Urology panel discussion on the selection criteria and intervention for fetal bladder outlet obstruction. *Journal of Pediatric Urology, 13*, 345–351 https://doi.org/10.1016/j.jpurol.2017.02.021.

Glick, P. L., Harrison, M. R., Golbus, M. S., Scott Adzick, N., Filly, R. A., Callen, P. W., ..., & deLorimier, A. A. (1985). Management of the fetus with congenital hydronephrosis II: prognostic criteria and selection for treatment. *Journal of Pediatric Surgery, 20*, 376–387 https://doi.org/10.1016/s0022-3468(85)80223-2.

Godinho, A. B., Nunes, C., Janeiro, M., Carvalho, R., Melo, M. A., & da Graça, L. M. (2013). Ureterocele: antenatal diagnosis and management. *Fetal Diagnosis and Therapy, 34*, 188–191 https://doi.org/10.1159/000353388.

Hansen, W. F., Cooper, C. S., & Yankowitz, J. (2002). Ureterocele causing anhydramnios successfully treated with percutaneous decompression. *Obstetrics & Gynecology, 99*, 953–956 https://doi.org/10.1097/00006250-200205001-00033.

Johnson, M. P., Corsi, P., Bradfield, W., Hume, R. F., Smith, C., Flake, A. W., ..., & Evans, M. I. (1995). Sequential urinalysis improves evaluation of fetal renal function in obstructive uropathy. *American Journal of Obstetrics and Gynecology, 173*, 59–65 https://doi.org/10.1016/0002-9378(95)90170-1.

Mandell, J., Peters, C. A., Estroff, J. A., & Benacerraf, B. R. (1992). Late onset severe oligohydramnios associated with genitourinary abnormalities. *The Journal of Urology, 148*(2 Pt 2), 515–518.

Montebruno, X. T., Torres Montebruno, X., Martinez, J. M., Eixarch, E., Gómez, O., García Aparicio, L., ..., & Gratacos, E. (2015). Fetoscopic laser surgery to decompress distal urethral obstruction caused by prolapsed ureterocele. *Ultrasound in Obstetrics & Gynecology, 46*, 623–626 https://doi.org/10.1002/uog.14876.

Morris, R. K., Malin, G. L., Quinlan-Jones, E., Middleton, L. J., Hemming, K., Burke, D., …, & Kilby, M. D. (2013). Percutaneous vesicoamniotic shunting versus conservative management for fetal lower urinary tract obstruction (PLUTO): a randomised trial. *The Lancet,*, *382*, 1496–1506 https://doi.org/10.1016/s0140-6736(13)60992-7.

Morris, R. K., Quinlan-Jones, E., Kilby, M. D., & Khan, K. S. (2007). Systematic review of accuracy of fetal urine analysis to predict poor postnatal renal function in cases of congenital urinary tract obstruction. *Prenatal Diagnosis*, *27*, 900–911 https://doi.org/10.1002/pd.1810.

Nassr, A. A., Shazly, S. A. M., Abdelmagied, A. M., Araujo Júnior, E., Tonni, G., Kilby, M. D., …, & Ruano, R. (2017). Effectiveness of vesicoamniotic shunt in fetuses with congenital lower urinary tract obstruction: an updated systematic review and meta-analysis. *Ultrasound in Obstetrics & Gynecology*, *49*, 696–703 https://doi.org/10.1002/uog.15988.

Persico, N., Berrettini, A., Fabietti, I., Dallagiovanna, C., Minoli, D., Mosca, F., …, & Manzoni, G. (2017). New minimally invasive technique for cystoscopic laser treatment of fetal ureterocele. *Ultrasound in Obstetrics & Gynecology: The Official Journal of the International Society of Ultrasound in Obstetrics and Gynecology*, *50*(1), 124–127.

Peters, C. A., Reid, L. M., Docimo, S., Luetic, T., Carr, M., Retik, A. B., …, & Mandell, J. (1991). The role of the kidney in lung growth and maturation in the setting of obstructive uropathy and oligohydramnios. *Journal of Urology*, *146*, 597–600 https://doi.org/10.1016/s0022-5347(17)37865-5.

Quintero, R. A., Homsy, Y., Bornick, P. W., Allen, M., & Johnson, P. K. (2001). In-utero treatment of fetal bladder outlet obstruction by a ureterocele. *The Lancet*, *357*, 1947–1948 https://doi.org/10.1016/s0140-6736(00)05084-4.

Quintero, R. A., Hume, R., Smith, C., Johnson, M. P., Cotton, D. B., Romero, R., …, & Evans, M. I. (1995). Percutaneous fetal cystoscopy and endoscopic fulguration of posterior urethral valves. *American Journal of Obstetrics and Gynecology*, *172*(1 Pt 1), 206–209.

Ruano, R., Sananes, N., Wilson, C., Au, J., Koh, C. J., Gargollo, P., …, & Belfort, M. A. (2016). Fetal lower urinary tract obstruction: proposal for standardized multidisciplinary prenatal management based on disease severity. *Ultrasound in Obstetrics & Gynecology*, *48*, 476–482 https://doi.org/10.1002/uog.15844.

Sananes, N., Favre, R., Koh, C. J., Zaloszyc, A., Braun, M. C., Roth, D. R., …, & Ruano, R. (2015). Urological fistulas after fetal cystoscopic laser ablation of posterior urethral valves: surgical technical aspects. *Ultrasound in Obstetrics & Gynecology*, *45*, 183–189 https://doi.org/10.1002/uog.13405.

Shankar, K. R., Vishwanath, N., & Rickwood, A. M. K. (2001). Outcome of patients with prenatally detected duplex system ureterocele; natural history of those managed expectantly. *Journal of Urology*, *165*, 1226–1228 https://doi.org/10.1016/s0022-5347(05)66494-4.

Shipp, T. D., Bromley, B., Pauker, S., Frigoletto, F. D., Jr., & Benacerraf, B. R. (1996). Outcome of singleton pregnancies with severe oligohydramnios in the second and third trimesters. *Ultrasound in Obstetrics & Gynecology: The Official Journal of the International Society of Ultrasound in Obstetrics and Gynecology*, 7(2), 108–113.

Smith-Harrison, L. I., Hougen, H. Y., Timberlake, M. D., & Corbett, S. T. (2015). Current applications of in utero intervention for lower urinary tract obstruction. *Journal of Pediatric Urology*, *11*, 341–347 https://doi.org/10.1016/j.jpurol.2015.07.012.

Soothill, P. W., Bartha, J. L., & Tizard, J. (2003). Ultrasound-guided laser treatment for fetal bladder outlet obstruction resulting from ureterocele. *American Journal of Obstetrics and Gynecology*, *188*, 1107–1108 https://doi.org/10.1067/mob.2003.179.

Spaggiari, E., Faure, G., Dreux, S., Czerkiewicz, I., Stirnemann, J. J., Guimiot, F., …, & Muller, F. (2017). Sequential fetal serum β2-microglobulin to predict postnatal renal function in bilateral or low urinary tract obstruction. *Ultrasound in Obstetrics & Gynecology: The Official Journal of the International Society of Ultrasound in Obstetrics and Gynecology*, *49*(5), 617–622.

Spaggiari, E., Stirnemann, J. J., Benedetti, S., Dreux, S., Salomon, L. J., Blanc, T., …, & Muller, F. (2018). Comparison of biochemical analysis of fetal serum and fetal urine in the prediction of postnatal renal outcome in lower urinary tract obstruction. *Prenatal Diagnosis, 38*, 555–560 https://doi.org/10.1002/pd.5263.

Tassis, B. M. G., Trespidi, L., Tirelli, A. S., Bocconi, L., Zoppini, C., & Nicolini, U. (1996). In fetuses with isolated hydronephrosis, urinary β2-microglobulin and n-acetyl-β-D-glucosaminidase (NAG) have a limited role in the prediction of postnatal renal function. *Prenatal Diagnosis, 16,* 1087–1094 https://doi.org/3.0.co;2-d">10.1002/(sici)1097-0223(199612)16:12<1087:aid-pd996>3.0.co;2-d.

CHAPTER 4

Diagnosis, clinical presentation, and radiologic assessment

Lisieux Eyer de Jesus[a], João Luiz Pippi-Salle[b]
[a]Pediatric Surgery and Urology, Antonio Pedro University Hospital; Servidores do Estado Federal Hospital, Rio de Janeiro, Brazil
[b]Sidra Medical and Research Center, Doha, Qatar

Clinical presentation and diagnosis

Ureteroceles are cystic dilatations of the submucosal terminal ureter that present anatomical variations, with several possible combinations of clinical and radiological presentations (uni or bilateral, orthotopic or ectopic, related to single or duplicated systems) (Fig. 4.1).

An important concept is that ectopic and extravesical ureteroceles are not synonyms: an ectopic ureterocele may be intravesical, as the definition of ectopic ureterocele refers to any ureterocele that does not correspond to the implantation of the distal ureter in the normal site (the lateral limit of the trigonal bar) (Merlini & Lelli-Chiesa, 2004). As a consequence, ureterocele inserted near the bladder neck, for example, is ectopic, but still intravesical. The misunderstanding of this concept may cause difficulties to compare research data about ureteroceles, due to equating intravesical to orthotopic ureteroceles or to excluding intravesical ureteroceles from analyses concerning ectopic ureteroceles.

There are two classifications for ureteroceles:

1. The ureteroceles are classified according to the anatomy into as orthotopic, ectopic, sphincteric, and ceco-ureteroceles (Glassberg et al., 1984).
2. The ureteroceles are classified according to anatomy and to the clinical expressions of the defect, including the presence of contralateral reflux or obstruction (Churchill et al., 1992).
3. The classification proposed by the Urology Section of the American Academy of Pediatrics, where the ureteroceles were didactically classified as single or duplex system and intravesical or extravesical. This one is the more frequently adopted contemporaneously, due to the conceptual frame that may be directly related to clinical consequences and embryological differences (Merlini & Lelli-Chiesa, 2004).

Ureteroceles
http://dx.doi.org/10.1016/B978-0-12-817888-1.00004-5

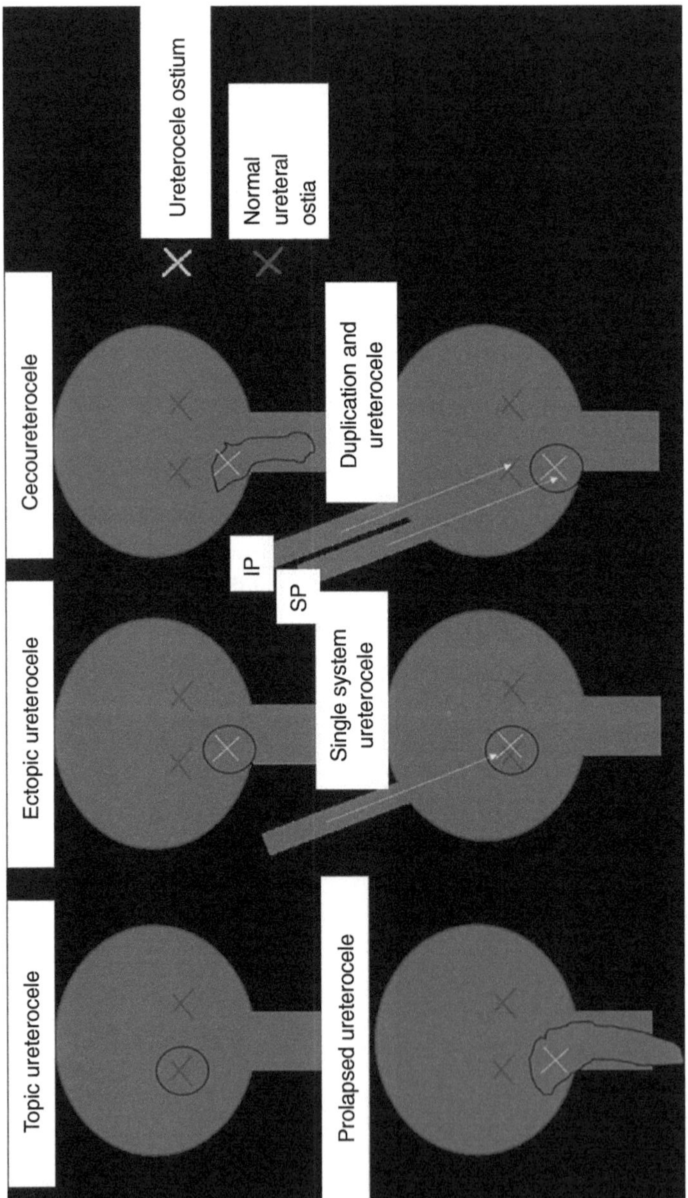

Figure 4.1 ***Anatomical types of ureterocele.***

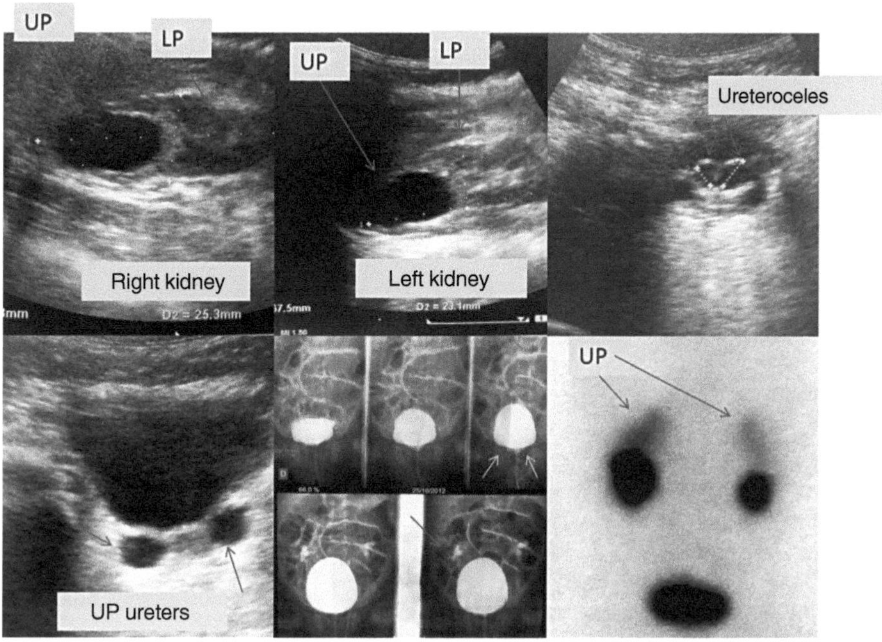

Figure 4.2 Bilateral ureterocele, ultrasound showing bilateral upper pole (UP hydronephrosis and cortical atrophy, dilated upper pole ureters (upper pictures and bottom left), cystourethrography showing bilateral lower pole reflux with "drooping lily" aspect (bottom middle) and hypofunctional upper poles (static scintigraphy, bottom right). *UP*, upper pole; *LP*, lower pole.

Duplex system ureteroceles (DSU) are the most frequent in Pediatrics (> three-fourth of the cases) (Andrioli et al. 2018) and are more frequent in females. DSU are always related to the caudal portion of the ureter that drains the superior pole (Weigert-Meyer embryological law), which is usually hydronephrotic and non-functional/hypofunctional and frequently dysplastic. The corresponding ureter is typically dilated and relates to a stenotic and ectopic ostium. In 10%–15% of the cases DSU are bilateral (Fig. 4.2).

The superior pole ureter may insert anywhere along the embryologic "ectopic pathway." In boys, this means the bladder, prostatic urethra, or genital ducts. In girls, an ectopic ureter may insert in the bladder, urethra, vulva (between the urethral meatus and the vaginal introitus), distal and upper vagina, uterus, and Fallopian tubes.

DSU are more frequently ectopic and intravesical (inserted distal and medial to the trigonal bar or in the bladder neck) or ectopic and extravesical (most commonly inserted in the bladder neck or urethra in girls).

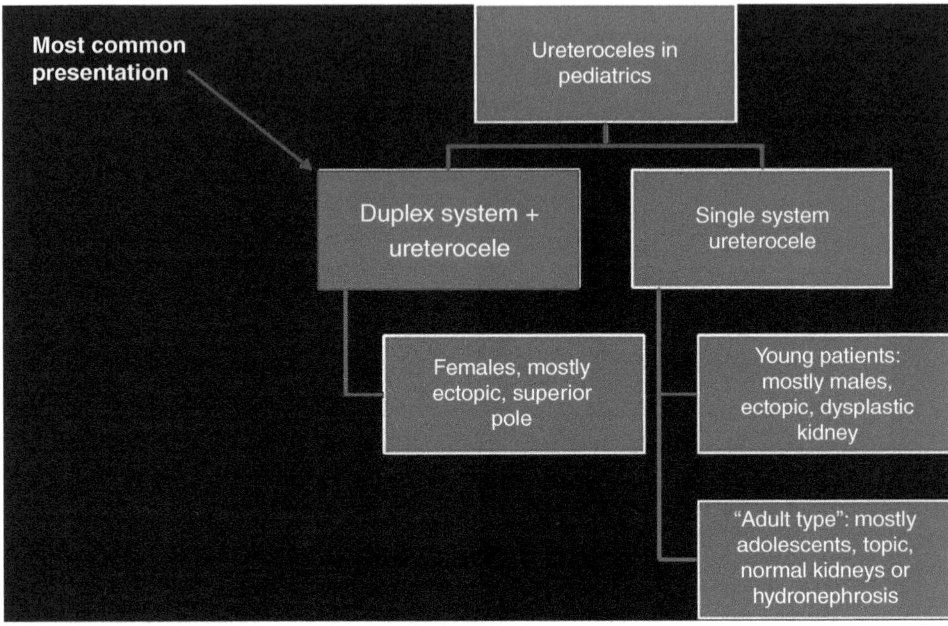

Figure 4.3 ***Ureteroceles in pediatrics.***

Orthotopic single system ureteroceles (SSU) are typical of adults and adolescents. They are frequently acquired and associated to mild/moderate hydronephrosis or to entirely normal renal units. In young children, SSU are congenital, usually ectopic/extra-vesical and related to non-functioning dysplastic kidneys. These are more frequent in males (Fig. 4.3).

Ureteroceles may be incidental findings in fetal evaluations or image exams. Exceptionally they cause fetal LUTO (lower urinary tract obstruction) syndrome.

The various anatomic forms relate to different clinical presentations that are modulated by the age and sex of the patient, by the anatomy and by the pressure inside the ureterocele (compressible, non-compressible and everting ureteroceles). In females, due to the short urethra, especially in young children, the ureteroceles may prolapse through the meatus and present as interlabial masses (Fig. 4.4). In males prolapsed ureteroceles are uncommon and located on the posterior urethra. They may present as LUTO related to a dilated posterior urethra, demanding differential diagnosis with posterior urethral valves (Grant & Bayne, 2018; Jesus & Franco, 2006).

Urinary tract infections (UTI) and urinary tract obstruction (UTO) are the final consequences of three main physiopathological phenomena, that may coexist (Fig. 4.5):

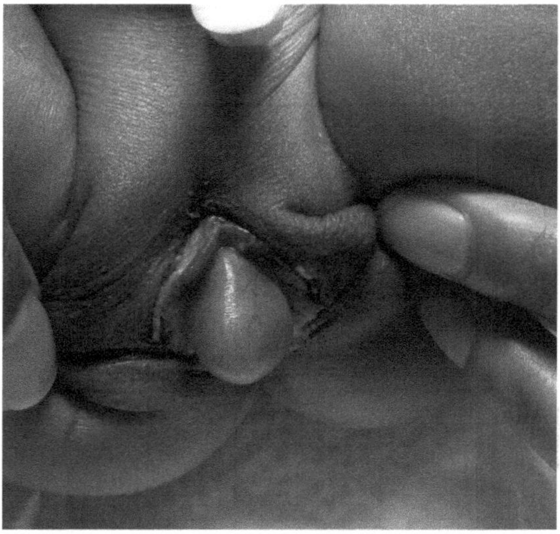

Figure 4.4 ***Female newborn with prolapsed ureterocele.***

1. Ureteral obstruction/stasis of the ureter from which the ureterocele originates, secondary to stenosis of the ureter/ureterocele termination.
2. Secondary obstruction of other anatomical structures:
   a. Ipsilateral lower pole ureter (in DSU), due to compression of the ostium/intravesical portion by the expanded ureterocele.
   b. Contralateral ureter: due to obstructed drainage secondary to compression/distortion by the expanded ureterocele.
   c. Bladder neck, due to direct blockage by the expanded ureterocele, submucosal invasion by the ureterocele (cecoureterocele) or filling by a prolapsed ureterocele.
   d. Urethra, secondary subepithelial invasion by the ureterocele (cecoureterocele) or filling by a prolapsed ureterocele.
3. Secondary VUR, by displacement of the lower pole ipsilateral ureter or bladder wall attenuation, caused by anatomic modification of the detrusor after ureterocele expansion and accommodation, especially in the trigonal area and in everting ureteroceles.

Prolapsed ureteroceles may show as negative images in the urethra in cystourethrographies (Fig. 4.6) or be directly shown by transperineal ultrasounds or, in young girls, as vulvar (interlabial) cystic masses after completely traversing the short urethra of the baby.

Prolapsed ureteroceles pass through the urethral canal, while cecoureteroceles represent an expansion of the ureterocele under the urethral mucosa (Fig. 4.7).

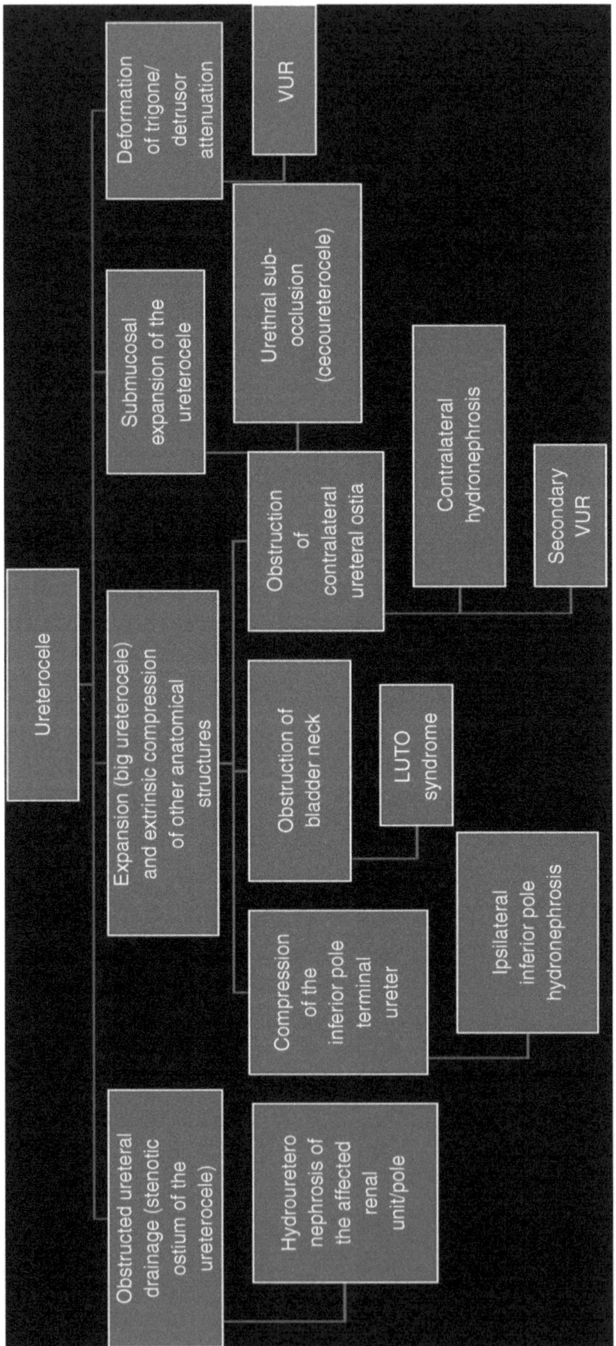

Figure 4.5 ***Ureteroceles physiopathology.***

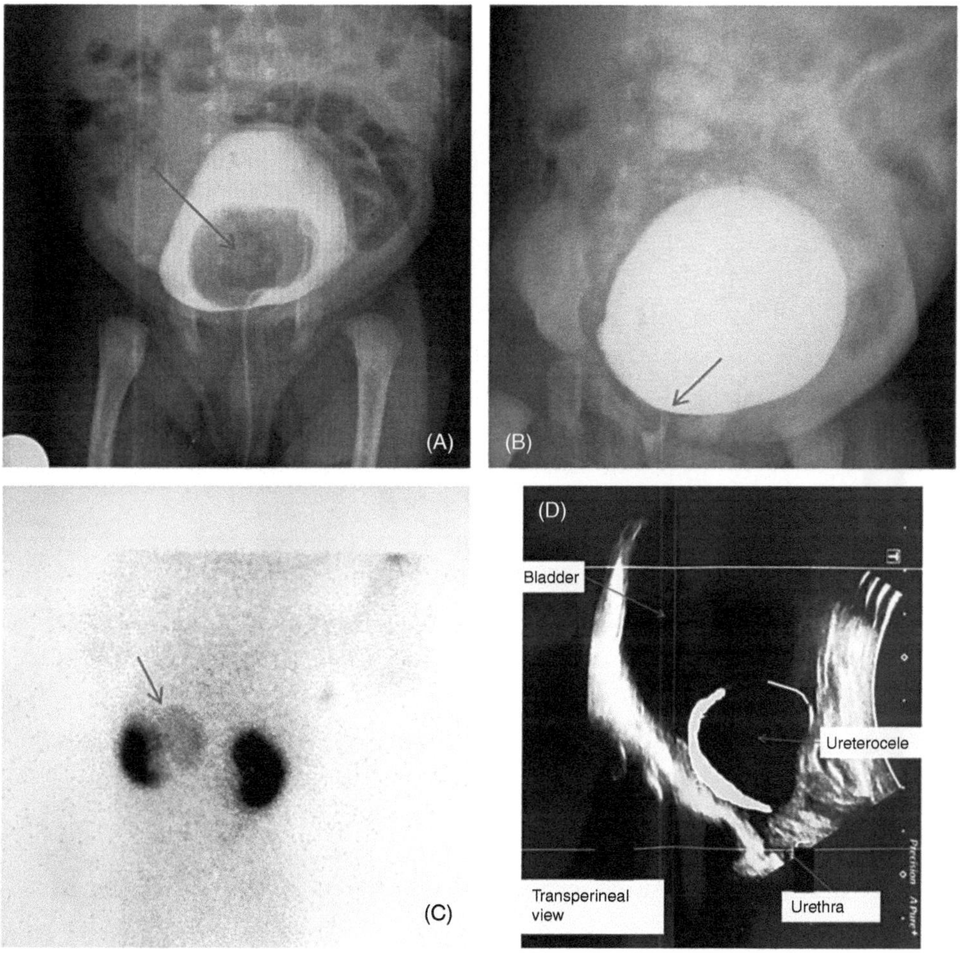

Figure 4.6 Large ureterocele, corresponding to a hydronephrotic and hypofunctional upper pole (*C*) shown in a contrast exam (cystourethrography) in two moments: intravesical in A and prolapsed in B *(arrows)*. The ureterocele is directly shown by transperineal ultrasound (*D*).

Exceptionally, ureteroceles in males may compound the clinical scenario in cases of Zinner syndrome (seminal vesical cysts, renal dysplasia/agenesis, ectopic ipsilateral ureter) (Maehana et al., 2018).

"Adult type" SSU may present with urinary calculi inside the ureterocele, but this is uncommon in children.

Dysfunctional symptoms may be present in patients with extra-vesical, ectopic ureteroceles, such as urinary incontinence or obstructed voiding

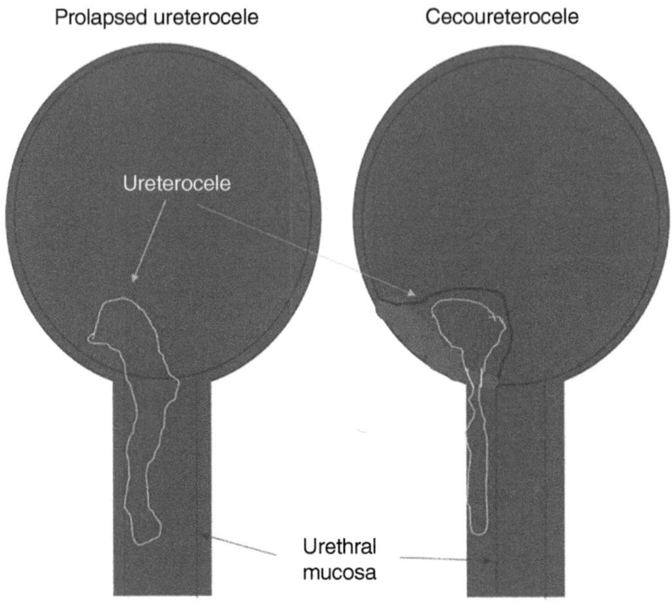

Figure 4.7 ***Differences between prolapsed ureterocele (left) and cecoureterocele (right).***

presentations (hesitation, flow abnormalities, straining to urinate, overflow with continuous dribbling of urine, stress incontinence, urinary retention, inability to toilet train) whenever the bladder neck is involved.

Voiding dysfunction may also present as new clinical complaints AFTER treatment of the ureterocele, unmasking stress incontinence after decompression of the ureterocele and resolution of obstruction or, alternatively, revealing iatrogenic complications (bladder neck injury or "windsock" behavior of residual cecoureteroceles) or showing functional sequelae of previous LUTO syndrome of bladder neck deformation. Bladder hyperactivity seems to be particularly common among girls and may need specific treatment.

Differential diagnosis

1. Bladder diverticula may be confounded with everting ureteroceles (5% of the ureterocele patients) (Didier et al., 2017).
2. Paravesical cysts, especially big prostatic utricles (in boys) (Shebel et al., 2013; Kavanaki et al., 2015) and Mullerian obstructions/hydrocolpos (in girls).

3. Zinner syndrome (unilateral renal agenesis or dysplasia associated to cysts of the seminal vesicle and ipsilateral ectopic ureter): The seminal vesicle abnormalities present as retrovesical cysts. Rarely, the syndrome may be associated to ureteroceles (Maehana et al., 2018). Patients may be asymptomatic (incidental finding) or (mainly after puberty) present dysuria, perineal pain, hemospermia, epididymitis, prostatitis, painful ejaculation, infertility, or UTI (Kavanaki et al., 2015).
4. Posterior urethral valves/LUTO syndrome, in case of prolapsed ureteroceles or ureteroceles obstructing the bladder neck in boys (Grant & Bayne, 2018; Jesus & Franco, 2006)
5. Interlabial masses, especially paraurethral (Skene) gland obstruction (Nussbaum & Lebowitz, 1983).

Radiologic assessment

Fetal ureteroceles

Ultrasound is the main exam used to diagnose fetal ureteroceles. In some cases fetal MRI may be indicated to get better anatomical information. Whenever ureterocele is detected, the examiner should look for:

1. Sex of the fetus.
2. Signs of renal duplication (DSU versus SSU).
3. Signs of renal dysplasia corresponding the renal segment drained by the ureterocele (hyperechogenicity, parenchymal cysts).
4. Ipsilateral and/or contralateral secondary hydronephrosis.
5. Signs of LUTO (thickened bladder wall, incomplete bladder emptying) and/or complications (oligohydramnios). Severe fetal LUTO secondary to ureteroceles is exceptional, but may determine fetal intervention (fetal puncture or drainage of the ureterocele) (Chalouhi et al., 2017; Godinho et al., 2013).
6. Description of the affected anatomical structures, description of extravesical portions of the ureterocele (ectopic or prolapsed ureteroceles).

Pediatric ureteroceles

Image (urinary tract ultrasound (UTUS), transperineal ultrasound, voiding cystourethrography (VCUG), intravenous pyelogram (IVP), retrograde pyelogram (Arevalo et al., 2017), CT, magnetic resonance urography (MRU), anatomical (cystoscopy) and functional (renal scintigraphy, dynamic MRU, urodynamics) exams are useful to:

1. Diagnose the malformation.
2. Diagnose the anatomical and functional consequences of the malformation to the urinary tract (hydronephrosis, VUR, voiding dysfunction, LUTO syndrome).
3. Diagnose associated renal dysfunction and/or dysplasia.
4. Determine the need and the method of treatment. Ureterocele treatment is imperative whenever the malformation is causing obstruction (urgent), UTI, voiding dysfunction, loss of function, worsening hydronephrosis and lithiasis. In the absence of those problems the patient may be managed conservatively, according to individualized evaluation. Despite this theoretical argument, most duplex system ureteroceles need surgery. Single system ureteroceles are frequently treated conservatively (Andrioli et al., 2018).
5. Diagnose complications of the ureterocele (UTI/pyelonephritis/renal sequelae/urolithiasis).
6. Follow up after treatment and diagnosis of new post-treatment VUR. De novo VUR is frequent after treatment of ureteroceles (occurs in a third to half of the cases), typically but not exclusively to the upper pole after endoscopic puncture of the ureterocele. New VUR to the lower pole, however also occurs or, alternatively, unmasks, supposedly due to morphological adaptations of the trigone after resolution of the ureterocele, which may also occur after heminephrectomies (Jesus et al., 2011).

UTUS is the most frequent exam to diagnose ureteroceles and their consequences, due to availability, low cost, absence of radiation, no risk of iatrogenic UTI, and acceptability by pediatric patients. Typical ureteroceles are seen as intravesical cystic masses with a regular and thin wall, occupying the trigonal region of the bladder (cyst within cyst), frequently extending near to the bladder neck and eventually to the urethra, in most cases associated to a dilated ureter (Fig. 4.4). Most are part of the clinical picture of DSU and signs of duplication are also evidenced in the exam. The duplex kidney may either be larger or shorter than the contralateral (in this last case due to hypotrophy of the superior pole) and may show asymmetric hydronephrosis or dysplasia limited to the superior pole. Two renal hila and/or a separation "bar" of renal parenchyma between the superior and the inferior poles may be seen. The exam is useful to detect signs of obstruction (secondary hydronephrosis, high post-void bladder residuals, and bladder wall thickening), as well as to infer the dynamics of the system by describing variations in the diameter of the ureterocele in various bladder volumes, anatomical limits of the ureterocele when fully expanded, eversion of the trigonal wall during voiding and indirect signs of vesicoureteral reflux (dynamic ultrasound). Especially in young babies, transperineal ultrasound may add useful information about ectopic ureteroceles (Jesus, Fazecas, Ribeiro & Dekermacher, 2017) (Fig. 4.5). Ruptured or punctured ureteroceles show as collapsed solid irregular structures related to the trigone. After endoscopic treatment or heminephrectomy/nephrectomy of the affected renal unit the ureterocele may show either as a collapsed peritrigonal structure or as smaller cysts. In the case of cecoureteroceles treated exclusively by endovesical endoscopic puncture, the ureterocele may present most of the time as an empty collapsed structure and fill with urine during voiding, eventually causing urethral obstruction (wind-sock phenomenon). In order to detect this problem by ultrasound one needs a dynamic evaluation, sometimes associated to a transperineal ultrasound.

Children presenting with ureteroceles should be submitted to a voiding cystourethrogram (VCUG)(Figs. 4.2, 4.6, and 4.8). The indication of this test to detect VUR is controversial, but not in the presence of a ureterocele. UTUS are sensitive to detect the ureteroceles per se but not VUR. Although the need to detect VUR in asymptomatic children is debatable, VCUGs are useful to determine the consequences of the ureterocele on the bladder anatomy and dynamics (bladder wall eversion during voiding, signs of bladder obstruction—trabeculation, pseudodiverticula, high post-void residual volume, obstruction, and deformity of the bladder neck and urethra).

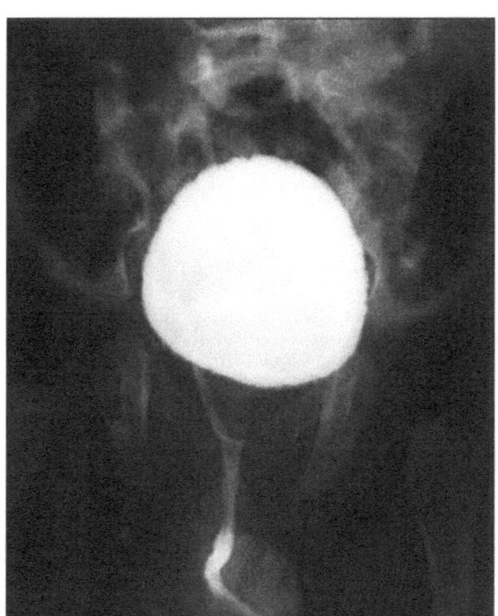

Figure 4.8 ***Prolapsed ureterocele to the posterior urethra in a male infant.***

Radionuclide cystography, indirect voiding scintigraphy and ultrasound with contrast agents are useful to follow up VUR, but are not favored as first choices in ureterocele patients, considering the need to depict anatomical details of the urinary tract anatomy.

Importantly, images of ureteroceles vary according to the degree of bladder repletion. A visualized ureterocele in early bladder filling and may not be seen after bladder repletion, as they are usually compressible by intravesical pressure. VUR is most commonly associated to the ipsilateral lower poles, but it can also occur in the contralateral kidney or ipsilateral upper pole after endoscopic puncture of the ureterocele. VUR to the lower poles characteristically show a lateral deviation of the ureter (the axis of the ureter points to the ipsilateral shoulder of the child, as compared to a non-duplicated ureter, which is approximately parallel to the vertebral column) (Didier et al., 2017) (Fig. 4.2).

IVP is not frequently used contemporaneously, due to better image definition in other exams, exposure to radiation and the need to use intravenous contrast (potentially nephrotoxic and allergenic) (Fig. 4.9). They are contraindicated in the first weeks of life, due to renal immaturity and low capacity of concentration, generating low quality images. On the other hand, the exam is easily available and relatively inexpensive, associated to lower levels of radiation as compared to CT, provides good images and doesn't need deep sedation or anesthesia (Plazas et al., 2011).

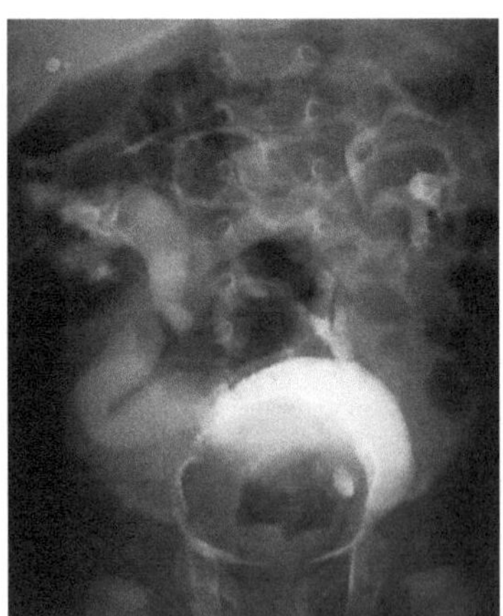

Figure 4.9 ***Intravenous pyelogram in a case of bilateral duplication, showing exclusion of the right upper pole, hydroureteronephrosis with lateralization of the lower pole ureter, two calices, and lower position of the of the right lower pole (drooping lily sign) and a huge intravesical ureterocele.***

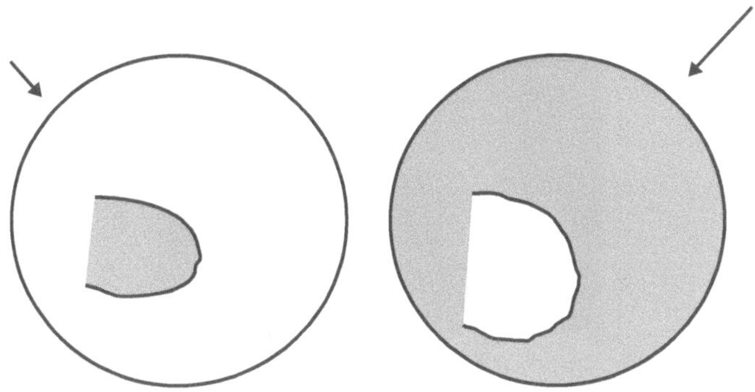

Figure 4.10 ***Distinction between "adult" and "infantile" ureteroceles in IVP.***

The classical "cobra head" image in IVP (showing the ureterocele as an intravesical projection full of contrast in the trigonal region) is typical of "adult type" SSU, relating to normally functioning kidneys. It is an uncommon finding in children: typical childhood ureteroceles are associated with non-functioning renal units. In Pediatrics the ureterocele is usually seen as a bladder intraluminal-filling defect with regular contour, corresponding to the trigonal region/bladder neck (Figs. 4.9 and 4.10).

A duplicated kidney with a non-functioning superior pole may seem shorter. The related renal unit may be indirectly evidenced by an inferior and lateral dislocation of the inferior pole (drooping lily sign) that usually shows a number of main calices that would be unusual to a normal kidney (typically two). The lower pole in duplications and the contralateral renal unit may show hydronephrosis or sequelae of UTI (chronic pyelonephritis) and the bladder may present signs of obstruction (trabeculation and pseudodiverticula).

Static renal scintigraphies (DMSA/SRS) are used to determine the function of the affected renal units and the presence of renal "scars" (signaling either sequelae of previous UTI/chronic pyelonephritis or dysplasia). Non-function is expressed as absent sectional parenchymal uptake of the radioisotope. Most authors describe superior poles showing ≤5%–10% of global function as hypofunctional (Andrioli et al., 2018; Faure et al., 2014).

A non-functional upper pole may appear as a "partial amputation" of the kidney on DMSA scans. In other cases the non-functional upper pole may be very small, without the typical indentation, misleading the investigator to diagnose a normal looking kidney. A high degree of suspicion to detect the abnormality is required in such cases (Fig. 4.11).

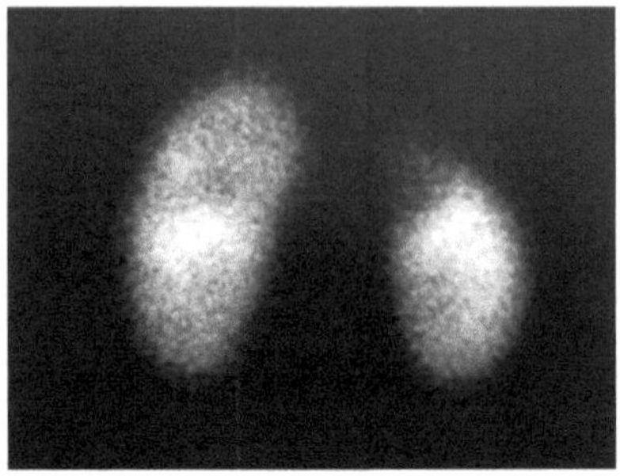

Figure 4.11 ***Static scintigraphy (DMSA), posterior view, showing hypofunctional (left) or "amputated" (right) right kidney upper pole.***

Whenever both units are normal in a duplicated kidney the static renal scintigraphy shows a larger than normal organ, but this situation is unusual among pediatric ureterocele cases. Secondary lower pole hydronephrosis appears in SCS as eccentric central photopenic areas.

Static renal scintigraphy is useful to determine the relative function of the various kidney units. In duplicated kidneys with normal function the relative function (summation of the function of the two units) should be >50%: a lower total function of the kidney suggests hypofunction of one or both units (Stokland et al., 2007). Ideally, the examiner should determine the function of each sub-unit separately, as the relative function of the affected superior pole may be a criterion to choose preservation of a functional superior unit versus heminephrectomy. A normal superior pole in a duplication is responsible for approximately 20% of the global function (Vates et al., 1996), but most authors consider that ≥5%–10% of relative function in a superior pole is enough to consider preservation of the segment, but this cut-off value is controversial and somewhat arbitrary (Faure et al., 2014).

Considering the high ionizing radiation doses associated with the exam, especially in multiphase protocols, CT is not the first choice to evaluate the urinary tract in ureterocele cases and is reserved for patients needing better demonstration of anatomical details after other image exams. Contemporaneously MRI may be preferable, albeit demanding deep sedation/general anesthesia. CT, however, provides excellent resolution and anatomic details. Key findings are precise anatomical evaluation, depiction of a parenchymal bar separating the superior and the inferior poles, asymmetric hydronephrosis affecting different kidney segments, signs of upper pole dysplasia and/or dysfunction, the presence of two ipsilateral ureters and the anatomy of the ureterocele.

MRU (magnetic resonance urography), especially contrast-enhanced dynamic MRU, is useful in ureterocele cases and may associate anatomic to functional and dynamic data without exposing the patient to ionizing radiation. The disadvantages are the need for anesthesia, high cost, relatively low availability, and being a time-consuming test. The urine in the urinary tract may serve as contrast, but dynamic information relies on the usage of gadolinium-based intravenous contrast and multiphase MRU. One advantage of the exam is the possibility of depicting detailed images of the ureters and of the ureterocele, providing useful morphological information even with a non-functional moiety, in sequences using urine as contrast. MRU has a superior role in demonstrating occult non-functional non-dilated upper poles and in depicting the insertion of ectopic ureters/ureteroceles, as compared to other image exams (Avni et al., 2001).

Information obtained from the diverse image exams is strongly correlated to irreversible severe histological lesions in resected specimens and may be used to justify ablative surgery. Severe cortical thinning (detected by ultrasound, CT, and MRI), high degree of hydronephrosis (detected by ultrasound, CT, and MRI) and poor function (detected by SRS, IVP, CT, and MRI) are strong predictors of irrecoverable function (Faure et al., 2014).

Voiding dysfunction is frequent in ureterocele patients. A higher frequency of primary bladder dysfunction is controversial. For some authors the incidence of voiding dysfunction among patients treated with endoscopic ureterocele puncture or kidney surgery is similar to that found in the normal population (Paye-Jaouen et al., 2015), but secondary dysfunction due to prolonged bladder neck obstruction, trigonal/bladder neck anatomical abnormalities or post-operative incontinence secondary to iatrogenic bladder neck or bladder nerve injury are possible. These patients may be evaluated with "non-invasive" urodynamics (UTUS, dynamic US, and flowmetry) (Filgueiras et al., 2003) to evaluate bladder wall thickness, post-void residual, abnormal/obstructed flow and bladder dynamic. Some patients (a minority) need a complete "classical" urodynamic evaluation, especially those presenting incontinence without obstructed voiding that are suspicious of bladder neck insufficiency and those needing detailed depiction of detrusor function.

Endoscopy (cystoscopy and vaginoscopy) is the most invasive modality to evaluate ureteroceles, but is useful to diagnose and also to treat ureteroceles. Endoscopy can exactly describe the anatomy of the ureterocele in relation to the bladder, lower pole and contralateral ureter, bladder neck and urethra or female genital tract, compressibility, potential to evert or to prolapse and the presence of deformities of the bladder neck or trigone. The exam needs to be done under a moderate degree of bladder distention, in order to allow good visibility, while not compressing the ureterocele. Cystoscopy may be used to decompress the ureter/bladder and resolve the urinary tract obstruction by puncture and/or incision of the ureterocele.

References

Andrioli, V., Guerra, L., Keays, M., Keefe, D. T., Tang, K., Sullivan, K. J., …, Garland, K., Rafikov, M., & Leonard, M. P. (2018). Active surveillance for antenatally detected ureteroceles: predictors of success. *Journal of Pediatric Urology, 14*, e1–e6.

Arevalo, M. K., Prieto, J. C., Cost, N., Nuss, G., Brown, B. J., & Baker, L. A. (2017). Utility of retrograde ureterocelogram in management of complex ureterocele. *Journal of Pediatric Urology, 13* e1-7.

Avni, F. E., Nicaise, N., Hall, M., Janssens, F., Collier, F., Matos, C., …, & Metens, T. (2001). The role of MR imaging for the assessment of complicated duplex kidneys in children: preliminary report. *Pediatric Radiology, 31*(4), 215–313.

Chalouhi, G. E., Morency, A. M., de Vleger, R., Martinez, G. M., Blanc, T., Hodges, R., …, Gueneuc, A., Ryan, G., deprest, J., Gratacos, E., & Ville, Y. (2017). Prenatal incision of ureterocele causing bladder outlet obstruction: a multicenter case series. *Prenatal Diagnosis, 37*, 968–974.

Churchill, B. M., Sheldon, C. A., & McLorie, G. A. (1992). The ectopic ureterocele: a proposed practical classification based on renal unit jeopardy. *Journal of Pediatric Surgery, 27*(4), 497–500.

Didier, A., Chow, S., Kwatra, N. S., etik, A. B., & Lebowitz, I. (2017). The duplicated collecting system of the urinary tract: embryology, imaging appearances and clinical considerations. *Pediatric Radiology, 47*, 1526–1538.

Faure, A., Merot, T., Sala, Q., haumoitre, K., Guys, J. M., Bourliere-Najean, B., …, Torrents, Mudler, O., Lechevallier, E., & Alessandrini, P. (2014). Value of diagnosis imaging in the evaluation of the severity of histological lesions in duplex systems. *Pediatric Urology, 10*, 361–367.

Filgueiras, M. T. F., Lima, E. M., Sanchez, T. M., Goulart, E. M. A., Menezes, A. C., & Pires, C. R. (2003). Bladder dysfunction: diagnosis with dynamic US. *Radiology, 227*, 340–344.

Glassberg, K. I., Braren, V., Duckett, J. W., Jacobs, E. C., King, L. R., Lebowitz, R. L., …, Permutter, A. D., & Stephens, F. D. (1984). Suggested terminology for duplex systems, ectopic ureters and ureteroceles. *Journal of Urology, 132*(6), 1153–1154.

Godinho, A. B., Nunes, C., Janeiro, M., Carvalho, R., Melo, M. A., & daGraça, L. M. (2013). Ureterocele: antenatal diagnosis and management. *Fetal Diagnosis and Therapy, 34*, 188–191.

Grant, C., & Bayne, C. (2018). Ureterocele causing bladder neck obstruction. *Journal of Pediatrics, 198*, 319.

Jesus, L. E., Farhat, A., Amarante, C. M., Dini, R. B., Leslie, B., Bagli, J., …, Lorenzo, A., & Pippi-Salle, J. L. (2011). Clinical evolution of vesicoureteral reflux following endoscopic puncture in children with duplex system ureteroceles. *Urology, 196*, 1455–1459.

Jesus, L. E., Fazecas, T., Ribeiro, B. G., & Dekermacher, S. (2017). Transperineal ultrasound as a tool to plan surgical strategies in pediatric urology: back to the future? *Urology, 104*, 175–178.

Jesus, L. E., & Franco, H. L. (2006). Prolapsed ureterocele in a small infant. *Pediatria (São Paulo), 28*, 59–64.

Kavanaki, A., Vidal, I., Merlini, L., & Hanquinet, S. (2015). Congenital seminal vesicle cyst and ipsilateral renal agenesis (Zinner syndrome): a rare association and its evolution from early childhood to adolescence. *European Journal of Surgery Reports, 3*, 98–102.

Maehana, T., Fukuta, F., Kobayashi, K., Hirobe, M., Tanaka, T., & Masumori, N. (2018). Laparoscopic surgery for seminal vesicle cysts and ureterocele with urination disorder: a case report of zinner syndrome. *Journal of Endourology Case Report, 4*, 35–38.

Merlini, E., & Lelli-Chiesa, P. (2004). Obstructive ureterocele, an ungoing challenge. *World Journal of Urology, 22*(2), 107–114 2004.

Nussbaum, A. R., & Lebowitz, R. L. (1983). Interlabial masses in little girls. *American Journal of Roentgenology, 141*, 65–71.

Paye-Jaouen, A., Pistolesi, F., Botto, N., Enezian, G., Grapin-Dagorno, C., Peycelon, M., ..., & Ghoneimi, A. (2015). Long-term bladder function after ureterocele decompression in children. *Journal of Urology, 193*(5S), 1754–1759.

Plazas, X. R., Ruiz, M. L., Rodriguez, D. A., Barranco, L. F., Velez, D. M., & Moragues, M. O. (2011). Ureterocele, intravenous urography images. *Archivos españoles de urología, 64*, 918–919.

Shebel, H. M., Farg, H. M., Kolokythas, O., & El-Diasty, T. (2013). Cysts of the lower male genitourinary tract: embryologic and anatomic considerations and differential diagnosis. *Radiographics, 33*, 1125–1143.

Stokland, E., Jodal, U., Sixt, R., Swerkersson, S., & Hansson, S. (2007). Uncomplicated duplex kidney and DMSA scintigraphy in children with urinary tract infection. *Pediatric Radiology, 37*(8), 826–828.

Vates, T. S., Bukowski, T., Triest, J., Friedman, A., Smith, C., Perlmutter, A., et al. (1996). Is there a best alternative to treating the obstructed upper pole? *Journal of Urology, 156*, 744–746.

CHAPTER 5

Active surveillance

Michael P. Leonard[a], Veridiana Andrioli[b]
[a]Children's Hospital of Eastern Ontario, University of Ottawa, Ottawa, ON, Canada
[b]Department of Surgery, Division of Urology, Federal University of Sao Paulo, Sao Paulo, Brazil

Introduction

The routine use of fetal ultrasound (US) as a screening tool during pregnancy has had considerable impact on pediatric urological care. In the era before antenatal sonography, significant urinary tract abnormalities would come to light by clinical presentations comprising urinary tract infection, flank pain, gross hematuria, urinary incontinence, and bladder outlet obstruction. Such presentations would lead to radiological investigation and targeted surgical intervention to address the anatomical problem. This reactive pattern of practice has now been supplanted by a preemptive model. Since we are now detecting urinary tract dilatation by ultrasound in approximately 1%–5% of fetuses (Ismaili et al., 2003), there is an opportunity to intervene to prevent future harm, particularly in the setting of obstructive uropathy. The vast majority of such planned intervention is performed in the infant, as antenatal intervention is wrought with controversy, and after 30 years is still considered experimental (Clayton & Brock, 2018). However, the main issue to be addressed is which of these radiological findings will become clinically significant, and who can be spared needless surgery. The pro-active practitioner is thus tasked with the conundrum of patient selection for surgical intervention versus careful follow-up over time to prevent harm in both instances.

Literature review

The first cohort in whom this selection process was studied comprised fetuses found to have significant unilateral hydronephrosis, which could be due to ureteropelvic junction obstruction (UPJO). Koff observed a cohort of 104 newborns noted to have antenatal hydronephrosis and found to have Society for Fetal Urology (SFU) grade 3–4 hydronephrosis on post-natal US (Fernbach, Maizels, & Conway, 1993). The indications for pyeloplasty comprised: >10% decrease in relative renal function on subsequent Lasix renography

Ureteroceles
http://dx.doi.org/10.1016/B978-0-12-817888-1.00005-7

studies or an increase in the degree of hydronephrosis on subsequent renal US. All of these infants were followed regardless of their initial degree of hydronephrosis, differential renal function, or renogram drainage curve. Of the cohort, 23/104 had a pyeloplasty at 3–14 months of follow up for the indications mentioned above. The remaining 76% were followed non-operatively and demonstrated a decrease in hydronephrosis over 2.5 years and stable or improved relative renal function, comparable to values seen in the surgical cohort (Ulman, Jayanthi, & Koff, 2000). This study was seminal as it showed that the radiological tools we have at hand are not necessarily predictive of the need for surgery, and that the majority of children with unilateral antenatal hydronephrosis which resembled possible UPJO did not require surgery. Thus there is a relatively benign natural history to unilateral antenatal hydronephrosis of this type, and many patients may be spared surgery.

The observation that antenatal hydronephrosis does not necessarily equate to the need for surgery was then applied to other antenatally detected conditions, including megaureters. In the absence of overt obstruction on renal scan or high-grade vesicoureteric reflux, the majority of such individuals have favorable outcomes without the need for surgery (DiRenzo et al., 2015). In those with high-grade hydronephrosis, antibiotic prophylaxis may be a useful adjunct to their management to reduce the risk of urinary tract infection in infancy (Easterbrook, Capolicchio, & Braga, 2017).

The concept of observing infants with antenatally detected ureteroceles was first brought forward by Rickwood from Liverpool. He reviewed ureterocele management at Alder Hey, and included in the 41 patients with duplex system ureteroceles were a small subgroup of 8 patients managed expectantly. These patients had no evidence of bladder outlet obstruction and a non-functioning upper pole moiety associated with the ureterocele, with the remainder of their upper tracts said to be normal. During follow-up ranging from 0.5 to 6 years (mean 2.3), none had symptoms and serial ultrasound exams were stable (Rickwood, Reiner, Jones, & Pournaras, 1992). The same group subsequently expanded their observational cohort to 14 patients, reported in a review of ureterocele management at their institution in 2001. It should be noted that the time interval during which the patients were accrued overlapped between their first and second publications, and they did not comment on whether some of the same patients were included twice. In any case, 14 children (8F/6M) with prenatally detected duplex system ureteroceles were followed for a median of 8 years (range 1.6–12.8). Patients were maintained on antibiotic prophylaxis until toilet trained or until age 5 years, if vesicoureteric reflux (VUR) was present. They were followed by serial imaging, which documented stable (8) or resolved (6) hydronephrosis and resolution ofVUR in 3/4. No patient developed a urinary tract infection or other relevant symptoms during follow-up. Patients they excluded from expectant management included those with "useful" upper moiety function (>10% on renal scan), obstruction of any renal unit or the bladder outlet, and high grade VUR (Shankar, Vishwanath, & Rickwood, 2001). Thus, Rickwood brought to the fore the concept that observation of patients with antenatally detected ureteroceles was possible, and that not all such patients require surgical intervention. Further studies were undertaken by others to see if the selection criterion for observation could be more finely tuned.

Coplen and Austin published their series of 8 infants who had antenatally detected ureteroceles associated with multicystic dysplasia. This constituted 25% of their antenatal ureterocele population and was comprised of half duplex and half single systems. There was no ureteric dilatation seen adjacent to the bladder on US in any of the patients. There were no cases of bladder outlet obstruction or high grade VUR that were included in the observational cohort. Renal scintigraphy documented absence of function associated with the dysplastic renal unit in all cases. With a mean follow-up of 36 months (range 14–54) only one patient developed a urinary tract infection, and none required surgery. Serial US documented the involution of the multicystic dysplastic renal parenchyma by 18 months in all (Coplen & Austin, 2004). This publication reinforced Rickwood's hypothesis that patients who have a renal unit of poor or minimal function associated with ureterocele, absent high-grade reflux and/or bladder outlet obstruction, can be safely observed. But what of those with adequate renal unit function in the face of ureterocele—can they also be observed?

In an attempt to answer this question, Rushton undertook to study a select cohort prospectively to establish whether Lasix renography could aid in selection of ideal candidates for observation. They prospectively evaluated 13 patients (11 F/2 M) with ureteroceles (11 duplex, 2 single systems) for expectant management. All patients were evaluated with RUS, VCUG, and Lasix renal scans. Of the cohort, 9 required no surgical intervention during a median follow up of 41 months (13–96). They noted that patients who avoided surgical intervention either had non-functioning renal units associated with the ureterocele, or adequate function of the renal unit with good drainage on renography. The four surgical interventions occurred for progressive obstruction (1) and breakthrough UTI (3). These were all patients with duplex systems with function and drainage on renography. However, the mean upper moiety relative to lower moiety differential function in the surgical group was less then that seen in those with duplex systems who were observed successfully (24.3% vs. 40.8%) and the half time of drainage was longer (median 12.5 minutes vs. 5.3 minutes) (Han et al., 2005). However, due to the small number of patients in this study, it was underpowered, and there was no statistical difference seen in relative upper pole renal function or drainage half time between the two groups. However, one can conclude that patients with nonfunctioning renal units associated with ureterocele do well with observation. There was also a trend favoring observation in patients with functioning renal units associated with ureteroceles that demonstrated good drainage on Lasix renal scan.

Following this, we published a review of a small number of patients with ureteroceles who were followed expectantly. Among this cohort (6 duplex, 4 single) the median follow up was 5 years (1–11). Patients were excluded from observation if they had bladder outlet obstruction or high-grade VUR on initial evaluation. They had to have either no function of the renal unit associated with the ureterocele, or adequate drainage in the setting of function. Patients were maintained on antibiotic prophylaxis for an average duration of 1.5 years. Hydronephrosis resolved in 6 individuals and reflux resolved in 2/4 with the condition. No patients had a UTI or the need for surgery during the follow-up interval (Direnna & Leonard, 2006).

The studies described earlier represent a mixed cohort of patients with variable study entrance assessments and inclusion/exclusion criteria. With the exception of one, they were retrospective reviews. Selection bias and small numbers of patients limit the conclusions regarding the safety of expectant management. However, there is some useful information that can be gleaned from them.

When it comes to the issue of safety, there are several potential risks patients could be exposed to if managed by active surveillance. The first obvious risk is that of UTI and possible urosepsis. If one aggregates the occurrence of UTI in the series reported earlier, 4/35 patients with duplex system ureteroceles experienced a UTI on antibiotic suppression within the duration of follow-up. Thus it has to be acknowledged that UTI is a risk and the parents require education as to how best to proceed if they think their child is affected. There was not enough detail in the papers to know how many patients required urgent endoscopic drainage, but awareness of the need for this intervention if the patient is not responding to antibiotic treatment is imperative for caregivers to have front of mind.

The next risk to be addressed is that of progression of hydronephrosis, which may imply worsening drainage of a functioning renal unit subtended by the ureterocele. Again with pooled data, it seems hydronephrosis resolved

(13/27) or remained stable (13/27) in the vast majority. Only one patient required surgery for progressive obstruction in the series by Rushton, as described earlier (Han et al., 2005). Thus, progression of hydronephrosis can occur, and would be considered an indication for surgical intervention.

What is the fate of VUR in the face of duplex system ureterocele patients under observation? Based on data from the quoted series, reflux resolved in 38%–71% of cases. Even dilated reflux (4/5) showed resolution. It is thought that breakthrough UTI is more common if reflux co-exists with ureterocele, and thus the recommendation for antibiotic prophylaxis until the VUR resolves, or the patient is of an age where the VUR is not a concern. The presence of VUR of any grade should therefore not preclude active surveillance of a patient with ureterocele.

Another concern, which may arise in an observational cohort, is the potential for loss of renal function in the affected renal unit. The real question is how useful are these renal units despite the fact they exhibit some function on renal scan? In Bolduc's series of upper pole hemi-nephrectomy specimens, severe irreversible dysplasia was seen in 69% overall, and the worst histology was associated with ectopic ureteroceles (Bolduc et al., 2002). Likewise, Abel showed that in 30 upper pole hemi-nephrectomy specimens, dysplasia (73%), and inflammation (50%) were commonplace, and these changes were not felt to be reversible (Abel, Lendon, & Gough, 1997). Furthermore, attempts at surgical preservation of the upper pole reveal an average change in relative renal function of the affected renal unit on renal scan of $-2.25 \pm 2.34\%$ (range -6 to 12). This was not found to be significantly different from a cohort in the same study that underwent upper pole hemi-nephrectomy (Vates et al., 1996). Thus it is unlikely that significant impairment of overall renal function will result from active surveillance.

If the dysplastic moiety associated with the ureterocele is left in situ, will it result in hypertension or even worse, malignancy? It is theoretically possible that a dysplastic upper moiety may produce excessive renin, resulting in renovascular hypertension. In a study by Levy where patients with preserved upper moieties in the setting of duplex system ureteroceles were compared to those who had hemi-nephrectomy, no statistical relationship was found between preservation of the upper moiety and hypertension over a median follow-up of 15 years (1–33) (Levy, Vandersteen, Morgenstern, & Husmann, 1997). As regards risk of malignancy, metanephric blastema has been seen histologically in hemi-nephrectomy specimens (Blew, Carpenter, & Leonard, 2002), but no reports of malignancy, particularly Wilms' tumor, have been forthcoming. Thus, it appears that leaving the dysplastic upper moiety in situ is not harmful. Of note, the remnant upper moiety is also at play in patients who have had endoscopic ureterocele incisions or upper tract preserving approaches such as ipsilateral ureteroureterostomy.

The risk of ureterocele prolapse is difficult to estimate, and can occur during the course of observation. In our most recently reported series (Andrioli et al., 2018), prolapse occurred in 2/58 females observed over a range of 7 days to 17.2 years. Both responded well to incision and reduction of the ureterocele. Clearly, parents of such children need to be aware of how this unpredictable complication presents, so that their children may receive appropriate urgent care.

Stone formation within the ureterocele may also be a concern. However, in the observational series reported to date, it has not occurred.

Based on the earlier information, one can come up with exclusion criteria for active surveillance that most clinicians would agree on. If a patient presents with non-resolving urosepsis, then clearly they require intervention, usually in the form of endoscopic drainage, and would not be candidates for observation. In addition, those patients with obstruction of a functional upper moiety subtended by ureterocele or bladder outlet obstruction should not be observed, but rather require surgical intervention. There may also be social or parental issues, which negate against observation as an option. For example, if a patient lives in a remote area of the country where access to medical care is not timely or has parents who seem unreliable, then active surveillance is not for them.

The potential benefits of active surveillance may thus be summarized as such: UTI's may occur, but most do not require emergent surgical intervention. Hydronephrosis usually diminishes or remains stable, with very few progressing and requiring surgery. Vesicoureteric reflux has the potential to resolve spontaneously, and even if it doesn't its presence in an older child may not be clinically relevant. Additionally, renal function is not significantly jeopardized with expectant management and the occurrence of hypertension is similar to a surgically managed cohort, and remains uncommon. Such complications as ureterocele prolapse and calculi can occur, but can be successfully managed in a minimally invasive manner.

The potential benefits of active surveillance need to be seen in light of the downside of such management. There is a significant follow up burden on the families, as frequent clinic visits with RUS and/or Lasix renal scans are imposed on them. The required duration of such follow up is unknown, and will likely require extension even into adulthood. The use of antibiotic prophylaxis for an uncertain duration is also an imposition. Usually, we discontinue such therapy once the child without VUR has toilet trained.

If VUR is present we generally continue the antibiotic until the VUR has resolved or the child is not at significant clinical risk despite the ongoing presence of VUR. Finally, acute crises such as urosepsis and prolapse may occur, and parents need to be educated as to clinical red flags, which would require presentation to an emergency department.

Recent data

Recently, we undertook a retrospective study of a large cohort of patients with ureteroceles who were managed expectantly (Andrioli et al., 2018). It was our hope that with a larger number for analysis, we could find clinical and radiological indicators of who would benefit most from observation with least likelihood of developing problems requiring surgical intervention. The aforementioned study comprised 102 patients. Of these, 78 (58 females and 20 males) had a duplex collecting system associated with an ipsilateral ureterocele, and were followed for 0.02–17.3 years. The remaining 24 (6 females and 18 males) had single system ureteroceles and were followed for 0.3–11.7 years. The overall median observation time was 1.2 years (range 0.7–3.1). The majority of the ureteroceles were antenatally detected, with 5 patients having ureteroceles found when they underwent abdominal US for non-urological reasons in infancy. In order to be eligible all patients had a confirmatory post-natal US as well as a VCUG. For both genders, inclusion criteria were deliberately based on imaging studies.

All patients were started on antibiotic prophylaxis, which in our opinion reduced the risk of UTI and controlled for both cohorts of male patients (circumcised and intact foreskin). Antibiotic prophylaxis was maintained until patients were fully toilet trained. Patients were excluded from observation if they had confirmed function and obstruction of the renal unit associated with the ureterocele on Lasix renal scan, or had radiological findings that indicated bladder outlet obstruction (BOO). Patients with other urological conditions, such as multicystic dysplastic kidney, neurogenic bladder, or children diagnosed with ureterocele following urinary tract infection were excluded from the study.

The main issue we hoped to resolve in this study was which group of patients would be best suited for active surveillance based on commonly available clinical and radiological parameters. To answer this, we looked for a cohort of patients that would be ideal for observation considering a low risk of breakthrough febrile UTI, and consequently, less likely to require surgical intervention.

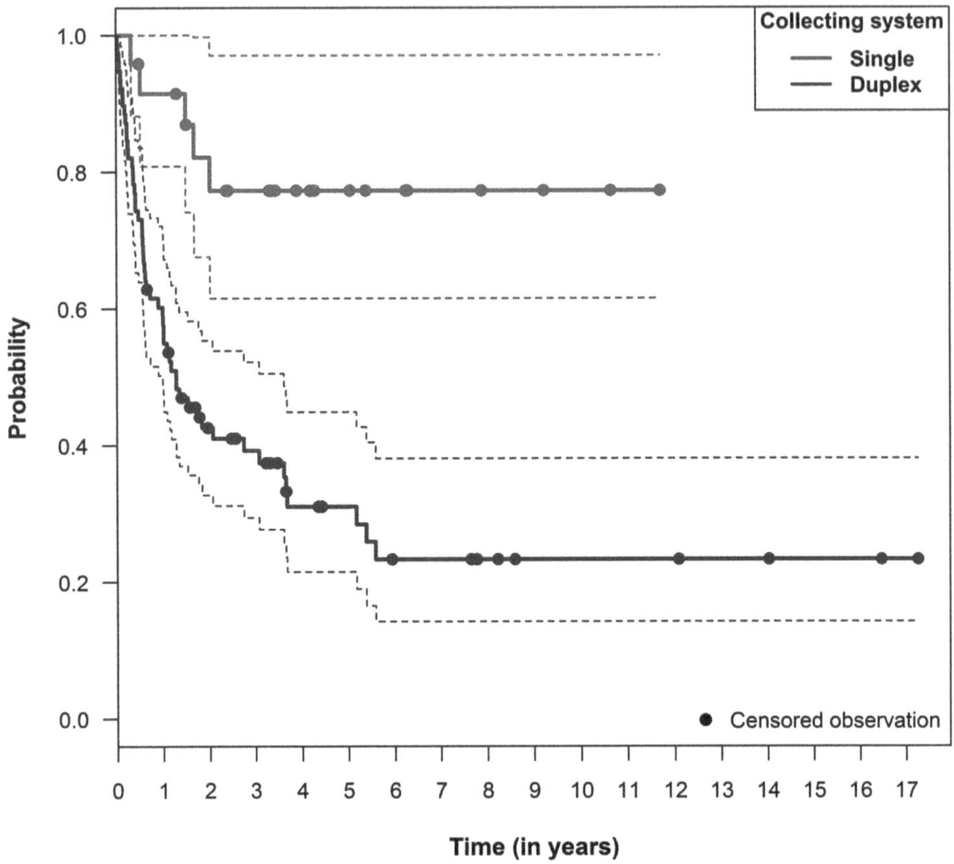

Figure 5.1 Kaplan- Meier curve comparing the probability of need for surgery in single versus duplex system ureteroceles. Single system patients are at low risk for surgery, and thus are ideal for observation.

The first important observation was that patients with single system ureteroceles are safer to observe than those with duplex systems. A Kaplan-Meier curve (Fig. 5.1) demonstrates a comparison between single and duplex system ureteroceles, and it shows that single system patients are safe to be followed conservatively as they are less likely to require surgery. Fully 80% of the single system ureterocele patients were followed without any surgical intervention. Our findings support previous publications as regards the safety of observing this particular group of patients (Feitz, Ritchey, & Bloom, 1994). Interestingly, all five children that required surgical intervention in the single system ureterocele cohort were male. One child had ureterocele incision after his follow-up US showed worsening hydronephrosis, three underwent nephrectomy for non-functioning kidneys after presenting with breakthrough febrile UTI's and one had ureteral reimplantation due to concurrent high grade vesicoureteric reflux.

As can be seen from Fig. 5.1, only 32% of patients with duplex system ureteroceles escaped the need for surgical intervention when placed on prophylactic antibiotics and surveillance. We feel this still represents a small victory for those who may in the past have been subjected to surgical intervention. In addition, since this was a retrospective study and incorporated the practices of several urologists over time, there may have been a bias toward surgical intervention, and if the study had been prospective, perhaps the survival on surveillance would have been greater for the duplex system cohort in the face of pre-established surgical indications.

That being said, we decided to analyze our patients with duplex system ureteroceles further to see if there were any indicators of those at risk versus those who would do well on active surveillance. We therefore first looked at the risk of developing a breakthrough febrile UTI in this cohort. Statistical analysis incorporated the following radiological findings: presence/absence of any grade of VUR, and hydronephrosis status defined as: (1) contralateral hydroureter or ipsilateral lower moiety hydronephrosis ± hydroureter, (2) contralateral hydronephrosis without hydroureter, and (3) absence of hydroureteronephrosis, excluding the status of the renal moiety associated with the ureterocele. We should highlight that all patients with duplex systems had some degree of hydronephrosis in the renal unit subtended by the ureterocele. A multivariable logistic regression model was then used to calculate predictive factors for breakthrough febrile UTI, incorporating three variables: gender, hydronephrosis and VUR. We did not find any gender predilection, and the most interesting finding was that VUR was not a factor that increased the risk of developing a breakthrough febrile UTI. The strongest predictor for breakthrough febrile UTI was the presence of hydronephrosis, specifically the presence of contralateral hydroureter or ipsilateral lower moiety hydronephrosis ± hydroureter (Odds Ratio (OR) 9.5, [1.2–71.7], $P = 0.028$). We did not specifically assess the influence of circumcision status versus the risk of developing a breakthrough febrile UTI. Given the retrospective nature of the study design and the small number of index events, an analysis in this realm could not be performed.

When surgical intervention in duplex systems was the focus of our statistical investigation, a Cox regression analysis was performed using gender, hydronephrosis status, VUR status and a history of breakthrough febrile UTI. This analysis showed that male gender (HR 1.8, [1.0–3.3], $P = 0.037$), or a history of fUTI (HR 3.1, [1.7–5.8], $P = 0.002$) increased the risk for surgical intervention. Given the retrospective nature of the study, we also found other drivers of surgical intervention since indications were not prospectively standardized. These included worsening hydronephrosis ($n = 15$), nonfunctioning upper moiety associated with the ureterocele ($n = 2$), prolapse ($n = 2$) and family decision ($n = 2$). Such indications speak to the possibility that had this study been prospective and indications for surgery decided ante tempus, that the survival on surveillance would have been higher for the duplex system cohort. The risk of prolapse should always be discussed with families of young girls being placed on surveillance so that it can be noticed and dealt with promptly as it was in both of our cases (Fig. 5.2).

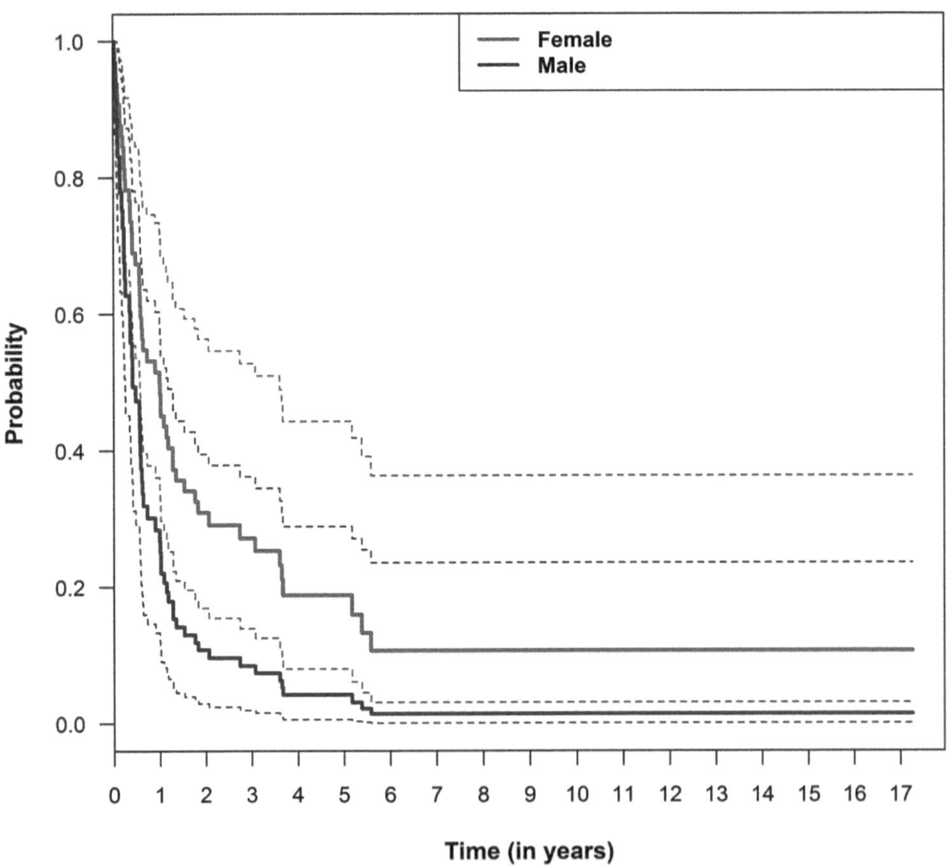

Figure 5.2 Male versus female duplex system ureterocele patients. Males with hydroureter (contra or ipsilateral) and ipsilateral hydronephrosis are more likely to fail conservative management.

Consolidating the earlier results, we concluded that for patients with duplex system ureteroceles, males with contralateral hydroureter and ipsilateral hydronephrosis ±hydroureter and a history of febrile urinary tract infection are more likely to fail active surveillance when compared to females with the same characteristics. It is obvious that the major influence on the need for surgical intervention was the occurrence of a breakthrough febrile UTI. However, three patients in the duplex system cohort had such an event and were not subjected to surgery. This again speaks to the lack of pre-ordained indications for surgical intervention, and the fragility of the data so generated.

Once again we have to highlight the pioneering work of Rickwood, Coplen and Austin, Rushton, and Leonard. These authors were the first to introduce the idea of active surveillance as a management option for ureteroceles. After them, many other authors published series of patients where surgical intervention might not be mandatory, each one emphasizing anatomic aspects that would be favorable for observation (Pohl, 2011) (Gander, Asensio, Royo, & Lloret, 2016). However, in all of these cases, evidence was limited and weakened by the retrospective design of each study and the small number of patients/events reported in each paper.

Ureterocele management is truly a "Pandora's box." An international survey done in 2010 where the authors obtained answers from 233 pediatric urologists, confirmed that there was no consensus regarding how they treat ureteroceles, except for the fact that they agreed to maintain patients on antibiotic prophylaxis (Merguerian, Taenzer, Knoerlein, McQuiston, & Herz, 2010). However, the most impressive fact was the fact that the majority of the professionals that answered this survey see less than 10 cases of ureteroceles per year. Therefore, it is not a common pathology, which creates some limitations in designing a standardized practice.

With the intention to enlighten physicians' decisions regarding ureteroceles, Table 5.1 below summarizes the radiological presentation that would select toward active surveillance in patients with antenatally detected ureteroceles. Of course, no clinical decision for observation should be taken without the complete education and engagement of the patient's parents and the provision of ready access to hospitals that can provide resources to resolve any complication that occurs in these children.

The improvement in imaging with antenatal US has provided an important tool for physicians to identify fetal malformations. They can then discuss with prospective parents the prognosis and possible interventions, which will be required. The interventions may take place in utero or shortly

Table 5.1 Radiological and clinical criteria favoring active surveillance for antenatally detected ureteroceles.

Anatomy, function	Authors
Single system	• Rushton (2005) • Leonard (2006) • Bloom (2016)
Non-functional upper pole (DMSA <10%) or MCDK	• Rickwood (2001) • Coplen and Austin (2004) • Rushton (2005) • Pohl (2011) • Lloret (2016)
No hydronephrosis	• Rickwood (2001) • El Ghoneimi (2009) • Leonard (2006 • Andrioli 2018
Absence of VUR or low grade VUR	• Rickwood (2001) • Leonard 2006 • El Ghoneimi (2009)
No evidence of obstruction on Lasix renogram	• Rushton (2005) • Leonard (2006)
Female	• Andrioli (2018)

after delivery. In utero intervention is now a reality in terms of spina bifida. It has led to a decreased requirement for ventriculo-peritoneal shunts, but has not resulted in an improvement in urological outcomes (Lee et al., 2012). The early experience with antenatally placed vesicoamniotic shunts was not particularly favorable and led to some pause in offering such intervention in fetuses presumed to have posterior urethral valves (Manning, Harrison, & Rodeck, 1986). However, there are anatomic malformations that can be carried through life without compromising the patient's health, but if not correctly identified and managed properly can cause harm. We also have to take in consideration that unnecessary surgery also can be harmful, particularly when performed in infants or newborns. There have been recent concerns regarding neuro-developmental outcomes in infants subjected to general anesthesia, so the decision for surgery should be carefully considered and indicated (Zhang et al., 2015). It is our contention that not all patients with ureteroceles will ultimately require surgery. Although the data are admittedly gathered from studies limited by differing parameters, small numbers and imperfect design, we feel there are some criteria, which may select those individuals at lower risk on active surveillance.

Conclusion

Based on the above-mentioned studies, the optimum patients for active surveillance are those of either gender with single system ureteroceles or ureteroceles subtending a multicystic dysplastic renal unit. Amongst the duplex system cohort, the optimum candidates are females with absence of hydronephrosis or hydroureter (aside from the renal unit associated with the ureterocele), a poorly functioning or non-obstructed functioning upper pole and no or low grade VUR. It would be desirable to generate data from a prospective study to provide a management algorithm and better choose who would benefit from active surveillance. Such a study would allow one to standardize selection criteria and events, which would eliminate a patient from surveillance. Due to the relative rarity of ureteroceles, a multicenter study would be necessary. Until such time the data described earlier is the best at hand.

References

Abel, C., Lendon, M., & Gough, D. C. (1997). Histology of the upper pole in complete urinary duplication—does it affect surgical management? *British Journal of Urology, 80*(4), 663–665.

Andrioli, V., Guerra, L., Keays, M., Keefe, D. T., Tang, K., Sullivan, K. J., …, & Leonard, M. P. (2018). Active surveillance for antenatally detected ureteroceles: predictors of success. *Journal of Pediatric Urology, 14*, 243 https://doi.org/10.1016/j.jpurol.2018.02.012.

Blew, B., Carpenter, B., & Leonard, M. P. (2002). Incidentally detected nephrogenic rests in the setting of congenital obstructive uropathy. *The Canadian Journal of Urology, 9*(4), 1595–1598.

Bolduc, S., Upadhyay, J., Sherman, C., Farhat, W., Bagli, D. J., McLorie, G. A., …, & El-Ghoneimi, A. (2002). Histology of upper pole is unaffected by prenatal diagnosis in duplex system ureteroceles. *The Journal of Urology, 168*(3), 1123–1126 https://doi.org/10.1097/01.ju.0000025866.15856.0f.

Clayton, D. B., & Brock, J. W. (2018). Current state of fetal intervention for lower urinary tract obstruction. *Current Urology Reports, 19*, 12 https://doi.org/10.1007/s11934-018-0760-9.

Coplen, D. E., & Austin, P. F. (2004). Outcome analysis of prenatally detected ureteroceles associated with multicystic dysplasia. *The Journal of Urology, 172*(4 Pt 2), 1637–1639 discussion 1639.

Direnna, T., & Leonard, M. P. (2006). Watchful waiting for prenatally detected ureteroceles. *The Journal of Urology, 175*(4), 1493–1495 discussion 1495. https://doi.org/10.1016/S0022-5347(05)00676-2.

DiRenzo, D., Persico, A., DiNicola, M., Silvaroli, S., Martino, G., & LelliChiesa, P. (2015). Conservative management of primary non-refluxing megaureter during the first year of life: a longitudinal observational study. *Journal of Pediatric Urology, 11*(4), 226.e1-6. https://doi.org/10.1016/j.jpurol.2015.05.007.

Easterbrook, B., Capolicchio, J. -P., & Braga, L. H. (2017). Antibiotic prophylaxis for prevention of urinary tract infections in prenatal hydronephrosis: an updated systematic

review. *Canadian Urological Association Journal*, *11*(1–2 Suppl. 1), S3–S11 https://doi.org/10.5489/cuaj.4384.
Feitz, W. F., Ritchey, M. L., & Bloom, D. A. (1994). Ureterocele associated with a single collecting system of the involved kidney. *Urology*, *43*(6), 849–851.
Fernbach, S. K., Maizels, M., & Conway, J. J. (1993). Ultrasound grading of hydronephrosis: introduction to the system used by the Society for Fetal Urology. *Pediatric Radiology*, *23*(6), 478–480.
Gander, R., Asensio, M., Royo, G. F., & Lloret, J. (2016). Evaluation of the initial treatment of ureteroceles. *Urology*, *89*, 113–117 https://doi.org/10.1016/j.urology.2015.11.025.
Han, M.Y., Gibbons, M. D., Belman, A. B., Pohl, H. G., Majd, M., & Rushton, H. G. (2005). Indications for nonoperative management of ureteroceles. *The Journal of Urology*, *174*(4 Pt 2), 1652–1656.
Ismaili, K., Hall, M., Donner, C., Thomas, D., Vermeylen, D., & Avni, F. E. (2003). Results of systematic screening for minor degrees of fetal renal pelvis dilatation in an unselected population. *American Journal of Obstetrics and Gynecology*, *188*, 242–246 https://doi.org/10.1067/mob.2003.81.
Lee, N. G., Gomez, P., Uberoi, V., Kokorowski, P. J., Khoshbin, S., Bauer, S. B., …, & Estrada, C. R. (2012). In utero closure of myelomeningocele does not improve lower urinary tract function. *The Journal of Urology*, *188*(4 Suppl.), 1567–1571 https://doi.org/10.1016/j.juro.2012.06.034.
Levy, J. B., Vandersteen, D. R., Morgenstern, B. Z., & Husmann, D. A. (1997). Hypertension after surgical management of renal duplication associated with an upper pole ureterocele. *The Journal of Urology*, *158*(3 Pt. 2), 1241–1244.
Manning, F. A., Harrison, M. R., & Rodeck, C. (1986). Catheter shunts for fetal hydronephrosis and hydrocephalus. *Report of the International Fetal Surgery Registry. The New England Journal of Medicine*, *315*(5), 336–340 https://doi.org/10.1056/NEJM198607313150532.
Merguerian, P. A., Taenzer, A., Knoerlein, K., McQuiston, L., & Herz, D. (2010). Variation in management of duplex system intravesical ureteroceles: a survey of pediatric urologists. *The Journal of Urology*, *184*(4 Suppl.), 1625–1630 https://doi.org/10.1016/j.juro.2010.03.113.
Pohl, H. G. (2011). Recent advances in the management of ureteroceles in infants and children: why less may be more. *Current Opinion in Urology*, *21*(4), 322–327 https://doi.org/10.1097/MOU.0b013e328346d455.
Rickwood, A. M., Reiner, I., Jones, M., & Pournaras, C. (1992). Current management of duplex-system ureteroceles: experience with 41 patients. *British Journal of Urology*, *70*(2), 196–200.
Shankar, K. R., Vishwanath, N., & Rickwood, A. M. (2001). Outcome of patients with prenatally detected duplex system ureterocele; natural history of those managed expectantly. *The Journal of Urology*, *165*(4), 1226–1228.
Ulman, I., Jayanthi, V. R., & Koff, S. A. (2000). The long-term followup of newborns with severe unilateral hydronephrosis initially treated nonoperatively. *The Journal of Urology*, *164*(3 Pt. 2), 1101–1105.
Vates, T. S., Bukowski, T., Triest, J., Freedman, A., Smith, C., Perlmutter, A., …, & Gonzalez, R. (1996). Is there a best alternative to treating the obstructed upper pole? *The Journal of Urology*, *156*(2 Pt. 2), 744–746.
Zhang, H., Du, L., Du, Z., Jiang, H., Han, D., & Li, Q. (2015). Association between childhood exposure to single general anesthesia and neurodevelopment: a systematic review and meta-analysis of cohort study. *Journal of Anesthesia*, *29*(5), 749–757 https://doi.org/10.1007/s00540-015-2030-z.

CHAPTER 6

Minimally invasive management of the ureterocele

Sameer Mittal[a], Douglas A. Canning[b]
[a]Pediatric Urology, The Children's Hospital of Philadelphia, Division of Urology, Perelman School of Medicine, University of Pennsylvania, Philadelphia, PA, United States
[b]Children's Hospital of Philadelphia; Perelman School of Medicine, University of Pennsylvania, Philadelphia, PA, United States

Introduction

The minimally invasive technique of endoscopic puncture, or transurethral incision (TUI) is our most popular initial management of ureteroceles. This approach is effective in relieving obstruction and is generally safe. The surgeon must consider intraoperative technical concerns as well as post-operative outcomes, such as iatrogenic VUR and the need for second procedure when counseling patients and their families. Puncture is ideal in infants with single system intravesical ureteroceles or in the setting of infection in the upper pole of a duplex system. The risk of displacement of the bladder outlet with ensuing vesicoureteral reflux is higher in those with ectopic thick walled ureteroceles. This chapter will focus on the techniques in performing a transurethral incision, follow-up and post-operative outcomes.

Transurethral incision of the ureterocele (TUI)

There are two goals of management (1) minimize subsequent risk of infection through decompression, (2) relieve the obstruction of the upper pole moiety. Because the upper pole usually represents no more than 15% of total renal function, preserving or improving function is not of primary importance. In an asymptomatic infant, endoscopic puncture is preferred, but in a septic child, percutaneous drainage of the upper pole may be more effective for temporary treatment (Coplen & Duckett, 1995).

Patient positioning, anesthesia, and equipment

We identify many of these infants with prenatal ultrasonography, so many are less than 6 months when undergoing TUI. We work closely with the anesthesiologist to mitigate any anesthetic concerns in positioning and

Ureteroceles
http://dx.doi.org/10.1016/B978-0-12-817888-1.00006-9

management. The child is placed in modified lithotomy position. We administer intravenous antibiotic prophylaxis based on previous culture data available or based on the local biogram (at CHOP we use cefazolin 20–30 mg/kg/dose). If using electrocautery, the child is grounded and we use sterile water intravesical irrigation.

Pediatric endoscopic equipment is essential in management. Typically, we use a 7.5-Fr to 10-Fr endoscope for infants. A larger resectoscope can be used for older children. A 3-Fr Bugbee or angle-tipped wire is used on a cutting setting high enough to perform a controlled clean incision, in general 20–30 W. In older children, we have also used a Collins knife electrode with a pediatric resectoscope successfully.

Cystoscopy

The goal of puncture is to provide enough of a drainage incision to provide for decompression while minimizing the risk of iatrogenic vesicoureteral reflux. Prior to any incision, in order to identify the location of the puncture, one must clearly delineate the anatomy. Cystoscopic evaluation of the ureterocele can be difficult due to the visual capabilities of the pediatric endoscope and the size of the ureterocele. Visualization of the entire ureterocele, its borders and its relation to surrounding structures such as the bladder neck and the bladder wall is essential. As a routine, the surgeon visualizes the contralateral ureteral orifices and in cases of duplex ureterocele, the ipsilateral lower pole ureteral orifice. We use irrigation fluid sparingly, particularly at first. Over aggressive bladder filling will flatten the ureterocele or even cause it to prolapse making subsequent incision difficult.

Ureterocele incision with electrocautery or laser

The ideal location of the incision is dependent on the wall thickness of the ureterocele, the wall thickness of the bladder, relative location of the bladder neck and the ureterocele extension into the urethra. The goal is to decompress the ureterocele relieving the obstruction with the additional goal of minimizing risk of vesicoureteral reflux into the upper pole. To assure a full thickness incision through the ureterocele wall and yet prevent distal migration of the incision site through the bladder neck, we tend to make our puncture adjacent to the lateral sidewall of the

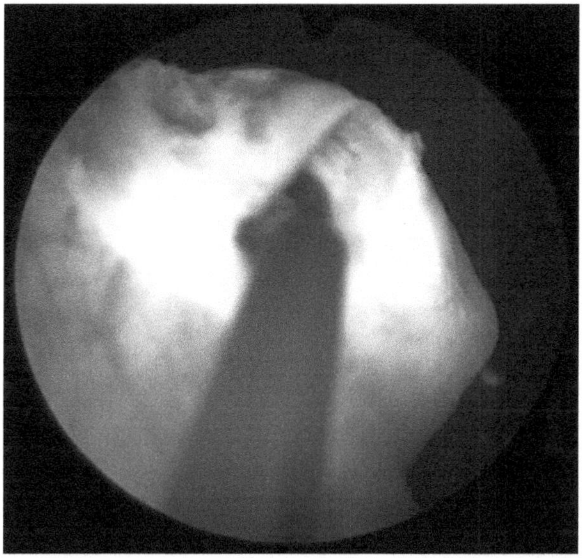

Figure 6.1 ***Cystoscopic image of transurethral incision with 3Fr angle-tipped wire.***

ureter just distal to the ureteral hiatus where the ureter passes through the bladder wall.

To assure a full-thickness incision into the ureterocele, we incise widely enough to allow clear visualization of the urothelium inside the ureterocele (Fig. 6.1). Once we have pinpointed the location of the proposed puncture, we incise boldly to prevent creation of a "false passage" that would displace the inner (ureterocele) urothelium away from the outer (bladder) urothelium resulting in difficulty maintaining orientation and an incomplete decompression of the obstructed ureter. At the same time, we take care to avoid perforation of the posterior wall of the bladder, which would result in risk of retroperitoneal or intraperitoneal urine leakage as well as possible bowel injury.

For an intravesical ureterocele, we make an incision inferior on the anterior lateral wall of the ureterocele to attempt to preserve an antireflux flap valve and collapse the ureterocele (Hagg et al., 2000) (Fig. 6.2A). Ectopic, cecoureteroceles or sphincteric ureteroceles are all managed with a single puncture of the intravesical portion of the lateral wall of the ureterocele well proximal to the bladder neck. This is achieved by following the sulcus with the bladder wall along the lateral edge of the ureterocele and then making a low puncture in the wall of the ureterocele (Fig. 6.2B). The portion of

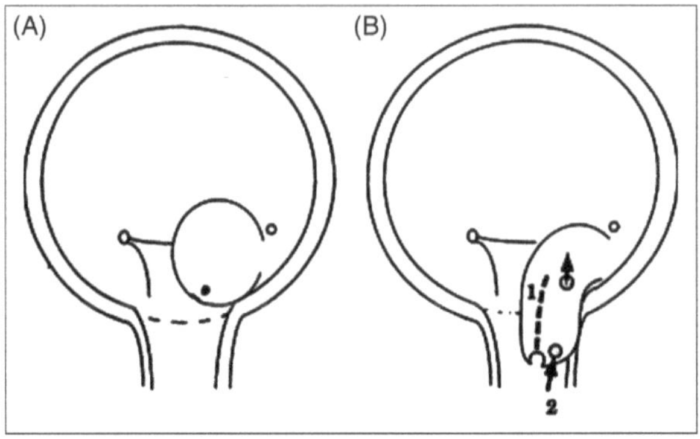

Figure 6.2 (A) An orthotopic ureterocele that will not project into the bladder neck with decompression may be incised as shown. (B) An ectopic ureterocele requires that the incision be made proximal to the bladder neck in a location that is superior enough to prevent the incision site from drifting into the bladder neck resulting in persistent upper pole obstruction. Incising the ureterocele relatively close to the lateral wall ensures that this will not occur. Alternatively, an incision from a position distal to the bladder neck superiorly to the bladder will also be effective. If this technique is employed, the surgeon must assure that a "flap valve" does not occur that can potentially obstruct the bladder outlet. *Adapted from Blyth et al., 1993.*

the incision superior to the bladder neck is important. Incisions distal to the bladder neck or incisions in a large ureterocele that are continued into the bladder neck as the ureterocele decompresses will be compressed by the internal sphincter, which will prevent adequate drainage.

As an alternative to traditional electrocautery, using a surgical laser has been reported to perform transurethral incision (Jankowski & Palmer, 2006; Marr & Skoog, 2002). A 200–550 μM laser fiber can be used with a holmium 2140 nM laser on settings of 0.5–1 J and 8–10 Hz to make an equivalent incision in the ureterocele that is performed with electrocautery. Authors have shown efficacy utilizing this modality with reported increased precision. Direct evidence comparing laser incision and electrocautery incision are lacking.

"Watering can" technique

The "watering can" ureterocele puncture was first described in the 1990s in Germany. The modern technique involves utilizing a 275–360 μM holmium laser fiber at a setting of 0.6 J and 6 Hz. The laser is used to make 10–20 puncture holes in the intravesical portion of the anterior and posterior wall of the ureterocele, creating a watering-can appearance (Fig. 6.3) (Palmer, Greger, Mannas, Kropp, & Frimberger, 2011).

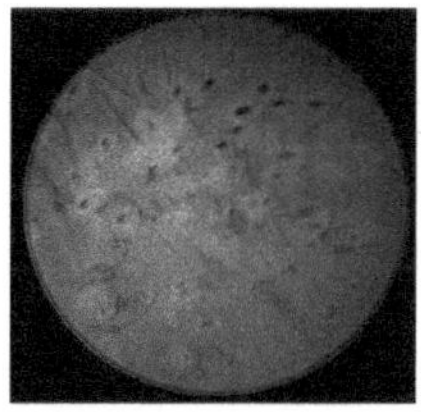

Figure 6.3 *"Watering-can" ureterocele puncture.*

Post-operative procedures and follow-up imaging

Endoscopic incision can be done effectively in an ambulatory surgery setting without the need for post-operative catheterization. It is our practice that if a patient is to be admitted for post-operative observation after anesthesia due to their age, a Foley catheter is left in place overnight and a trial of void is performed the following morning. We usually prescribe antibiotic prophylaxis following the procedure, but this is difficult to justify if there has been no previous history of urinary tract infections.

In general, a repeat renal and bladder ultrasound is performed 4–6 weeks after the incision to ensure reduction in the degree of renal dilation. A voiding cystourethrogram is performed 6 months after surgery to identify if VUR is present. If VUR is not noted, we discontinue the antibiotic prophylaxis. We then order an ultrasound of the kidneys, ureters, and bladder after toilet training or if a UTI occurs. If worsening hydronephrosis, recurrent UTIs or incomplete bladder emptying is noted, a repeat voiding cystourethrogram (VCUG) and/or a bladder dysfunction work-up is performed.

Outcomes

Despite the low risk and the relative ease of performing a TUI, we still debate when this procedure is preferred. Although the rate of improvement in obstruction and kidney function is high, the subsequent development of postoperative VUR and its management, the need for secondary procedure and voiding dysfunction has led to attempts to risk-stratify patients undergoing TUI. In general, the child's trigonal anatomy and distortions thereof, drive the need for secondary procedures more than the initial approach to decompressing the ureterocele. The management of post TUI VUR is addressed in Chapter 10.

Early reports of TUI

Successful decompression of the ureterocele reportedly occurs in more than 90% of infants with preservation of function in more than 95% for single-system intravesical ureteroceles undergoing TUI (Blyth, Passerini-Glazel,

Camuffo, Snyder, & Duckett, 1993; Rich, Keating, Snyder, & Duckett, 1990; Smith et al., 1994). This success rate declines in extravesical ureteroceles to 75% and 50%, respectively. Additionally, the rate of postoperative VUR between intravesical and extravesical ureteroceles was reported to be 18% versus 47% and the subsequent requirement for secondary procedure was significantly higher in the extravesical group. This initial data and its subsequent longer-term follow-up (Cooper et al., 2000) has led us to conclude that initial endoscopic puncture is preferred in the single-system intravesical ureterocele. Controversy remains on the appropriate management of extravesical or ectopic, duplex, refluxing and in ureteroceles that subtend a poorly functioning upper pole. In most cases, we incise these systems initially and then individualize care subsequently (Coplen & Duckett, 1995).

Postoperative vesicoureteral reflux

VUR after TUI may occur in the upper pole ureter or in the associated lower pole ureter in the duplex system (Adorisio et al., 2011; Jesus et al., 2011). VUR occurs following TUI in 30%–60% of patients. Development of ipsilateral lower pole VUR after upper pole nephrectomy has also been reported frequently (Husmann et al., 1999). Spontaneous resolution of VUR associated with ureteroceles does occur, but at a lower rate than those seen with primary VUR. The management of postoperative VUR is discussed in Chapter 10.

"Watering-Can" technique versus standard incision and VUR

The "watering can" technique has been reported as a method for attaining equivalent efficacy while decreasing the rate of VUR and secondary procedure (Palmer et al., 2011). Haddad et al. compared 34 patients undergoing the "watering-can" technique and 21 patients undergoing standard TUI and showed no difference in ureterocele decompression, but a reduction in the rate of de novo VUR (32% vs. 67%; $P < 0.05$) and subsequent surgery due to de novo VUR (38% vs. 71%; $P < 0.05$). Additionally, the volume of de novo VUR seemed lower in the "watering can" group compared to standard TUI. A recent report with a modification of this technique using minimal fenestrations with only 4–8 discrete puncture sites on the anterior portion of the ureterocele has shown similar findings (Caione et al., 2019).

Need for other procedures

Although early reports on the subject of secondary procedures have been published, investigators performed a comprehensive meta-analysis to help identify risk factors leading to secondary procedures (Byun & Merguerian, 2006). Ectopic ureteral location (RR 2.78; 95% CI 2.06–3.74), duplex renal systems (RR 3.93; 95% CI 1.41–10.93) and preoperative reflux (RR 1.56; 95% CI 1.24–1.96) were all associated with significantly greater reoperation rates. These factors have been implicated in other reports as well (Di Renzo, Ellsworth, Caldamone, & Chiesa, 2010). An interesting conclusion of the meta-analysis showed that patients with two or more risk factors had no higher reoperation rate compared to those with only one risk factor. The authors suggest that each of those risk factors may be proxies for trigonal distortion rather than independent risk factors. Trigonal distortion from a large or thick walled ureterocele likely best predicts a secondary procedure or bladder neck reconstruction.

Bladder dysfunction after ureterocele incision

Bladder dysfunction is common in children born with ureteroceles. Following TUI or reconstruction, children with ureteroceles may develop infrequent or incomplete voiding, and overflow incontinence (Abrahamsson, Hansson, Sillen, Hermansson, & Hjalmas, 1998). A number of investigators have studied the role various surgical interventions play on the development of bladder dysfunction (Holmes, Coplen, Strand, Husmann, & Baskin, 2002; Paye-Jaouen et al., 2015). Holmes et al. showed an overall bladder dysfunction rate of 6% of patients with ureteroceles, all with ectopic ureteroceles of duplex systems. More importantly, they showed similar rates of dysfunction for patients treated with lower tract only, upper tract only or a combination of upper and lower tract surgery. Again, this data supports the notion that congenital trigonal distortion that is seen with ectopic ureteroceles plays a much larger role in the subsequent morbidity associated with the disease.

Conclusion

TUI is a safe procedure with low periprocedure morbidity for the initial management of ureteroceles. In some cases TUI is the only surgery required. Variations in TUI technique and energy modality utilized have been reported with similar efficacy. Decreased rates of VUR and secondary procedure with the "watering can" technique have been reported. Regardless of technique, specific patient and anatomical factors, such as an ectopic ureteral location, duplex renal systems, and pre-operative VUR are associated with increased de novo VUR and the need for secondary procedure and may be markers for underlying trigonal distortion that may be the most important predictor for the need for additional procedures.

References

Abrahamsson, K., Hansson, E., Sillen, U., Hermansson, G., & Hjalmas, K. (1998). Bladder dysfunction: an integral part of the ectopic ureterocele complex. *Journal of Urology, 160*(4), 1468–1470 doi: 10.1016/s0022-5347(01)62593-x.

Adorisio, O., Elia, A., Landi, L., Taverna, M., Malvasio, V., & Danti, A. D. (2011). Effectiveness of primary endoscopic incision in treatment of ectopic ureterocele associated with duplex system. *Urology*, 77(1), 191–194 doi: 10.1016/j.urology.2010.02.061.

Blyth, B., Passerini-Glazel, G., Camuffo, C., Snyder, H. M., 3rd., & Duckett, J. W. (1993). Endoscopic incision of ureteroceles: intravesical versus ectopic. *Journal of Urology, 149*(3), 556–559 discussion 560. doi:10.1016/s0022-5347(17)36146-3.

Byun, E., & Merguerian, P. A. (2006). A meta-analysis of surgical practice patterns in the endoscopic management of ureteroceles. *Journal of Urology, 176*(4 Pt. 2), 1871–1877 discussion 1877. doi:10.1016/S0022-5347(06)00601-X.

Caione, P., Gerocarni Nappo, S., Collura, G., Matarazzo, E., Bada, M., Del Prete, L., …, & Capozza, N. (2019). Minimally invasive laser treatment of ureterocele. *Frontiers in Pediatrics*, 7, 106 doi:10.3389/fped.2019.00106.

Cooper, C. S., Passerini-Glazel, G., Hutcheson, J. C., Iafrate, M., Camuffo, C., Milani, C., & Snyder, H. M., III (2000). Long-term followup of endoscopic incision of ureteroceles: intravesical versus extravesical. *Journal of Urology, 164*(3 Pt. 2), 1097–1099 discussion 1099–1100. doi:10.1097/00005392-200009020-00045.

Coplen, D. E., & Duckett, J. W. (1995). The modern approach to ureteroceles. *Journal of Urology, 153*(1), 166–171 doi:10.1097/00005392-199501000-00068.

Di Renzo, D., Ellsworth, P. I., Caldamone, A. A., & Chiesa, P. L. (2010). Transurethral puncture for ureterocele-which factors dictate outcomes? *Journal of Urology, 184*(4 Suppl.), 1620–1624 doi:10.1016/jjuro.2010.04.023.

Haddad, J., Meenakshi-Sundaram, B., Rademaker, N., Greger, H., Aston, C., Palmer, B. W., …, & Frimberger, D. (2017). "Watering can" ureterocele puncture technique leads to decreased rates of de novo vesicoureteral reflux and subsequent surgery with durable results. *Urology, 108*, 161–165 doi:10.1016/j.urology.2017.06.008.

Hagg, M. J., Mourachov, P. V., Snyder, H. M., Canning, D. A., Kennedy, W. A., Zderic, S. A., & Duckett, J. W. (2000). The modern endoscopic approach to ureterocele. *Journal of Urology, 163*(3), 940–943.

Holmes, N. M., Coplen, D. E., Strand, W., Husmann, D., & Baskin, L. S. (2002). Is bladder dysfunction and incontinence associated with ureteroceles congenital or acquired? *Journal of Urology, 168*(2), 718–719.

Husmann, D., Strand, B., Ewalt, D., Clement, M., Kramer, S., & Allen, T. (1999). Management of ectopic ureterocele associated with renal duplication: a comparison of partial nephrectomy and endoscopic decompression. *Journal of Urology, 162*(4), 1406–1409.

Jankowski, J. T., & Palmer, J. S. (2006). Holmium: yttrium-aluminum-garnet laser puncture of ureteroceles in neonatal period. *Urology, 68*(1), 179–181 doi:10.1016/j.urology.2006.01.066.

Jesus, L. E., Farhat, W. A., Amarante, A. C., Dini, R. B., Leslie, B., Bagli, D. J., …, & Pippi Salle, J. L. (2011). Clinical evolution of vesicoureteral reflux following endoscopic puncture in children with duplex system ureteroceles. *Journal of Urology, 186*(4), 1455–1458 doi: 10.1016/j.juro.2011.05.057.

Marr, L., & Skoog, S. J. (2002). Laser incision of ureterocele in the pediatric patient. *Journal of Urology, 167*(1), 280–282.

Palmer, B. W., Greger, H., Mannas, D. B., Kropp, B. P., & Frimberger, D. (2011). Comparison of endoscopic ureterocele decompression techniques. Preliminary experience—is the watering can puncture superior? *Journal of Urology, 186*(4 Suppl), 1700–1703 doi:10.1016/j.juro.2011.04.007.

Paye-Jaouen, A., Pistolesi, F., Botto, N., Enezian, G., Grapin-Dagorno, C., Peycelon, M., & El-Ghoneimi, A. (2015). Long-term bladder function after ureterocele decompression in children. *Journal of Urology, 193*(5 Suppl), 1754–1759 doi:10.1016/j.juro.2014.10.115.

Rich, M. A., Keating, M. A., Snyder, H. M., III, & Duckett, J. W. (1990). Low transurethral incision of single system intravesical ureteroceles in children. *Journal of Urology, 144*(1), 120–121 doi:10.1016/s0022-5347(17)39387-4.

Smith, C., Gosalbez, R., Parrott, T. S., Woodard, J. R., Broecker, B., & Massad, C. (1994). Transurethral puncture of ectopic ureteroceles in neonates and infants. *Journal of Urology, 152*(6 Pt 1), 2110–2112 doi:10.1016/s0022-5347(17)32332-7.

CHAPTER 7

Minimally invasive interventions—laparoscopic ureteric clipping

Rodrigo Romao[a], Armando J. Lorenzo[b]
[a]Departments of Surgery and Urology IWK Health Centre, Dalhousie University, Halifax, NS, Canada
[b]Pediatric Urology, Women's Auxiliary Chair in Urology and Regenerative Medicine, Hospital for Sick Children; Division of Urology, Department of Surgery, University of Toronto, Toronto, ON, Canada

Introduction

The surgical management of ureteroceles associated with a non-functioning upper moiety is marked by a wide range of alternatives. While reported outcomes for upper and lower urinary tract approaches are similarly acceptable, a unique set of morbidities can arise from either choice.

Upper urinary tract surgery, that is, heminephrectomy of a non-functioning upper moiety, can compromise the blood supply to the lower pole leading to loss of the renal unit. Although the reported incidence of this complication is low (<10% of cases), its occurrence is undeniably devastating and has been systematically described in the literature for both open and minimally invasive techniques (Gundeti, Ransley, Duffy, Cuckow, & Wilcox, 2005; Jayram et al., 2011). Furthermore, the durability of the upper urinary tract approach can also be called into question, as the need for secondary surgical procedures to the lower urinary tract is common, taking place in up to 20% of cases mainly due to infections and residual vesicoureteral reflux (VUR) (Gran, Kropp, Cheng, & Kropp, 2005).

Lower urinary tract approaches also carry the theoretical risk of compromising function of the lower renal unit when uretero-ureterostomy (UU) is performed. Extensive reconstructive procedures involving the bladder (ureterocelectomy ± bladder neck reconstruction and ureteral reimplantation) can be associated with increased perioperative morbidity and long-term complications such as lower urinary tract symptoms, voiding dysfunction and vesico-vaginal fistula.

Clipping of the distal ureter associated with a non-functioning upper moiety was initially described in the management of primary urinary incontinence related to an ectopic ureter in females (Romão et al., 2014). The

Ureteroceles
http://dx.doi.org/10.1016/B978-0-12-817888-1.00007-0

impetus to offer this counterintuitive approach (i.e., permanent obstruction of a system already affected by pathologic obstruction) was borne out of the renal transplantation literature.

In 1985, Baquero et al. reported the results of simple ligation of the native ureter without concurrent nephrectomy associated with pyeloureterostomy as a feasible urinary reconstruction method for patients undergoing renal transplantation (Baquero et al., 1985). The efficacy and safety of this procedure in transplant patients have been further substantiated by other authors in over 400 patients, both children and adults (Gallentine & Wright, 2002; Timsit et al., 2010).

After a positive initial experience treating females with primary urinary incontinence associated with an ectopic ureter, where nine patients experienced complete symptom resolution without subsequent episodes of urinary tract infections or flank pain (Romão et al., 2014), the authors expanded the indication of this technique to include patients with ureteroceles.

Technique

The outpatient procedure starts with a cystoscopy. Under ideal circumstances, the lower pole ureteric orifice is recognized and cannulated with a ureteric catheter to assist in identification during the laparoscopic stage of the surgery. Such cannulation may not be feasible, particularly with large ureteroceles and those causing significant anatomic distortion of the trigone.

A standard laparoscopic approach is utilized to identify the ureters at the level of the pelvic brim. The ureter associated with the ureterocele is usually dilated secondary to obstruction, which facilitates distinction from the lower moiety one. The sheath containing the two ureters is incised and the ureters are isolated using blunt dissection. The ureter associated with the ureterocele is individualized. Complete or partial transection of the ureter allows for aspiration of the ureterocele as well as the dilated upper moiety; complete decompression of the ureterocele can be documented intraoperatively with ultrasound. Alternatively, such decompression can be achieved with suction through an angiocath inserted percutaneously into the abdominal cavity and the ureter under direct vision. Transection of the ureter and aspiration of the ureterocele can also be achieved with a small inguinal incision used as a port site, through which the ureter is delivered after being isolated laparoscopically (Fig. 7.1). The ureter is dissected as far distal as possible and a laparoscopic clip is applied. A laparoscopic Endoloop® is an option for ligation after complete transection of the ureter (Lopes, Mello,

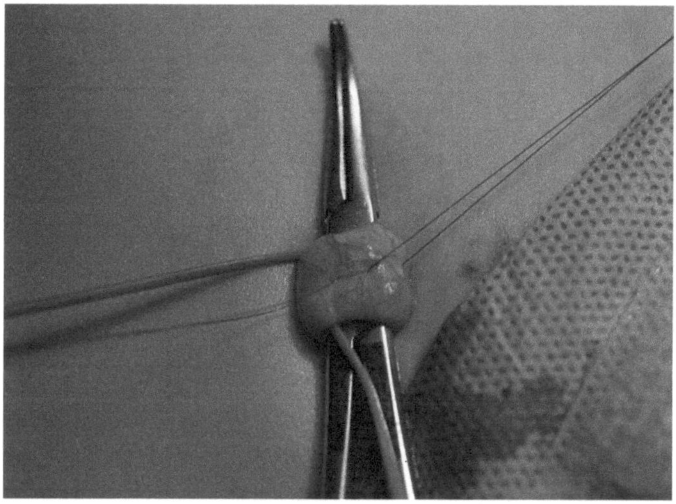

Figure 7.1 ***Dilated upper pole ureter delivered through a low abdominal incision used as a port site.***

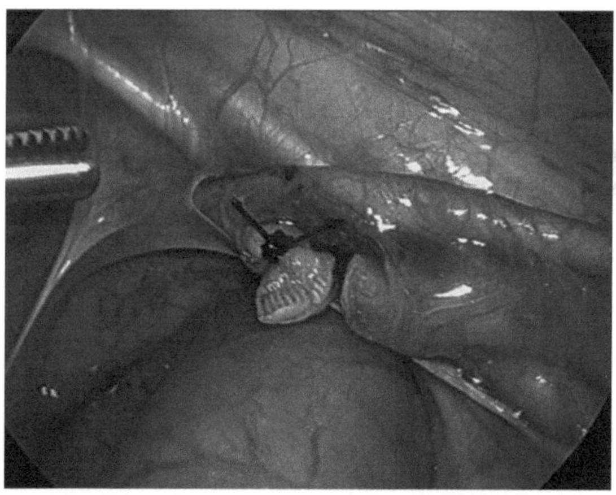

Figure 7.2 ***Appearance of upper pole ureter after laparoscopic ligation; note normal lower pole ureter containing ureteric catheter inserted cystoscopically right next to it.***

Koyle, & Lorenzo, 2019). Fig. 7.2 illustrates the final appearance after ligation of the upper pole ureter as low as possible.

The technique is appealing for its simplicity. The lower moiety and its respective collecting system are left undisturbed, provided that the identification of the ureters is performed adequately. There is no dissection in and around the renal pedicle and no risk for a urinary leak; there is no risk for de novo VUR, as observed with endoscopic ureterocele puncture or incision. The procedure is quick, simple and associated with minimal postoperative morbidity, no bladder spasms or need for indwelling catheters/stents. The patient can be discharged on the day of surgery and resume activities in a short period of time.

Clipping does not burn bridges, in the sense that should a complication arise such as severe infection or flank pain, the classic upper and lower urinary tract surgical options remain feasible and available to the surgeon.

Indications

Ureteric clipping should obviously only be considered for patients with minimal to non-functioning renal units. In patients where considerable upper pole residual function seems to be present on the renogram, a renal unit-sparing alternative (i.e., UU or bladder reconstructive procedure) should be entertained.

In patients where the indication for surgery is recurrent UTIs, it is important to ensure sterility of the urinary tract preoperatively and adequate intraoperative decompression prior to clipping. Irrigation with antibiotics may be an option in such cases, although not evidence-based.

Published experience with this technique is limited to less than 50 cases of duplex kidneys, of which only a few patients actually had a ureterocele. Nonetheless, excellent decompression of the ureterocele with minimal change in upper pole hydronephrosis (Figs. 7.3–7.5) without associated symptoms has been documented (Lopes et al., 2018, 2019).

As with any novel surgical procedure, close follow-up of patients undergoing ureteric clipping is still warranted to ensure no complications or symptoms arise over time.

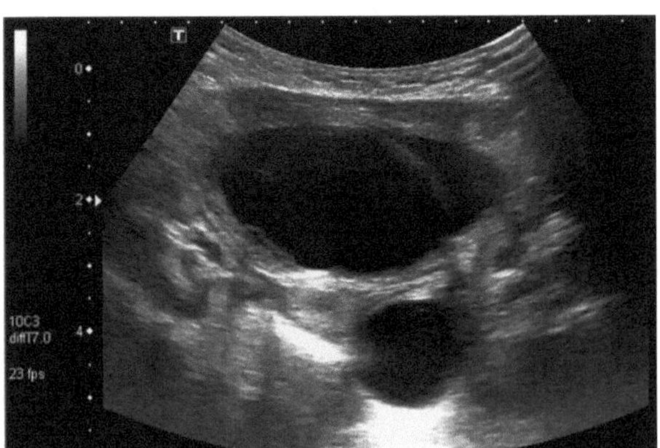

Figure 7.3 ***Large dilated upper pole ureter associated with ureterocele.***

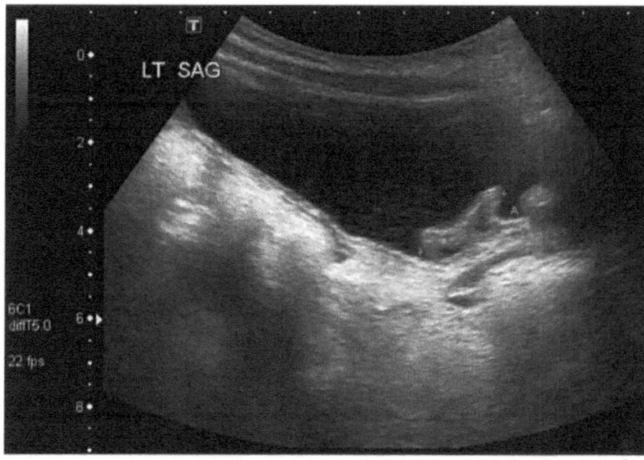

Figure 7.4 ***Decompressed ureterocele after ureteric clipping/ligation.***

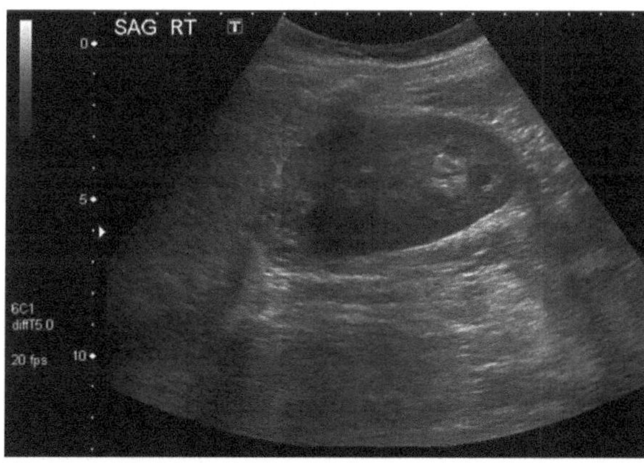

Figure 7.5 ***Appearance of the kidney after ureteric clipping with no evidence of significant upper pole hydronephrosis.***

References

Baquero, A., Ginsberg, P. C., Kaschak, D., Raja, R., Schneeberg, A., Solish, L., & Bannett, A. D. (1985). Experience with pyeloureterostomy associated with simple ligation of native ureter without ipsilateral nephrectomy in renal transplantation. *The Journal of Urology*, *133*(3), 386–387.

Gallentine, M. L., & Wright, F. H. (2002). Ligation of the native ureter in renal transplantation. *The Journal of Urology*, *167*(1), 29–30.

Gran, C. D., Kropp, B. P., Cheng, E. Y., & Kropp, K. A. (2005). Primary lower urinary tract reconstruction for nonfunctioning renal moieties associated with obstructing ureteroceles. *The Journal of Urology*, *173*(1), 198–201 http://doi.org/10.1097/01.ju.0000148374.64478.b5.

Gundeti, M. S., Ransley, P. G., Duffy, P. G., Cuckow, P. M., & Wilcox, D. T. (2005). Renal outcome following heminephrectomy for duplex kidney. *The Journal of Urology*, *173*(5), 1743–1744 http://doi.org/10.1097/01.ju.0000154163.67420.4d.

Jayram, G., Roberts, J., Hernandez, A., Heloury, Y., Manoharan, S., Godbole, P., et al. (2011). Outcomes and fate of the remnant moiety following laparoscopic heminephrectomy for duplex kidney: a multicenter review. *Journal of Pediatric Urology*, 7(3), 272–275 http://doi.org/10.1016/j.jpurol.2011.02.02.

Lopes, R. I., Fernández, N., Koyle, M. A., Santos, dos, J., Pérez, J., Romao, R. L., & Lorenzo, A. J. (2018). Clinical outcomes of the upper urinary tract after ureteral clipping for treatment of low functioning or nonfunctioning renal moieties. *Journal of Urology*, *199*(2), 558–564 http://doi.org/10.1016/j.juro.2017.09.080.

Lopes, R. I., Mello, M. F., Koyle, M. A., & Lorenzo, A. J. (2019). Ureteral clipping for the treatment of a non-functioning upper kidney moiety associated with a massive ureterocele: step-by-step description of a novel technique. *Journal of Pediatric Urology*, *15*(3), 284–285 http://doi.org/10.1016/j.jpurol.2019.03.005.

Romão, R. L. P., Figueroa, V., Pippi Salle, J. L., Koyle, M. A., Bägli, D. J., & Lorenzo, A. J. (2014). Laparoscopic ureteral ligation (clipping): a novel, simple procedure for pediatric urinary incontinence due to ectopic ureters associated with non-functioning upper pole renal moieties. *Journal of Pediatric Urology*, *10*(6), 1089–1094 http://doi.org/10.1016/j.jpurol.2014.04.008.

Timsit, M. -O., Lalloué, F., Bayramov, A., Taylor, M., Billaut, C., Legendre, C., et al. (2010). Should routine pyeloureterostomy be advocated in adult kidney transplantation? A prospective study of 283 recipients. *The Journal of Urology*, *184*(5), 2043–2048 http://doi.org/10.1016/j.juro.2010.06.144.

CHAPTER 8

Upper pole hemi-nephrectomy

Natalia Ballesteros[a,b], Rafael Gosalbez[b]
[a]Division of Pediatric Urology, Nicklaus Children's Hospital, Miami, FL, United States
[b]Department of Urology, University of Miami, Miami, FL, United States

Overview

In this chapter, the surgical management of ectopic ureteroceles is discussed with an emphasis on the associated controversies between the simplified and combined approaches. Techniques to the different modalities for upper pole hemi-nephrectomy (UHN) are described.

Introduction

Although the surgical management of pediatric ureteroceles can be challenging and controversial, the treatment goals have remained constant: to protect the lower pole (LP) moiety and preserve its renal function; to eliminate infection, obstruction or vesicoureteral reflux (VUR); and to preserve urinary continence (Gander, Asensio, Royo & Lloret, 2016).

The initial management of ureteroceles in duplex kidneys depends on multiple factors that include the age of the patient, the initial presentation, the associated ureterocele (intravesical or ectopic), the split function of the upper and lower moieties, and whether VUR is present. Of the surgical approaches, no single one can be considered curative for all patients and this adds to the unsolved debate as to which one is better. The current trend has moved toward less aggressive interventions (Gander et al., 2016; Mariyappa, Barker, Samnakay & Khosa, 2014). In some instances, a more conservative approach such as observation of asymptomatic patients or transurethral incision vs. puncture of a ureterocele may be performed first and potentially avoid the need for more invasive surgery. Endoscopic intervention is also indicated in emergencies such as sepsis in an infected obstructing ureterocele or when there is bladder outlet obstruction from a ureterocele prolapsing into the urethra. Indications for more invasive surgical interventions include persistent severe uretero-hydronephrosis, high grade dilating VUR or recurrent UTIs after endoscopic ureterocele management, and urinary incontinence secondary to an ectopic ureterocele (El-Ghoneimi

Ureteroceles
http://dx.doi.org/10.1016/B978-0-12-817888-1.00008-2

et al., 2003; Singh, Wagener & Chandran, 2010). Ureteroureterostomy and ureteropyelostomy, which are frequently done for obstructing ureteroceles with good-functioning UPs, have gained popularity in the last few years as an alternative to UHN in poor-functioning UP systems (Michaud & Akhavan, 2017). These surgical interventions act to decompress the ureterocele resulting in restoration of the trigonal anatomy and may also result in the resolution of existing VUR in some patients (Husmann, Ewalt, Glenski & Bernier, 1995). More recently, authors have reported just clipping the ureter of the poorly functioning or non-functioning upper pole moiety (UP) as a simpler and quicker surgical approach, which has shown good short-term outcomes. The role of this technical modification still needs to be determined in the management of these patients (Lopes et al., 2018; Sharifi, Nabavizadeh, Mozafarpour, & Kajbafzadeh, 2017). However, in many instances the residual ureterocele and/or ureteral stump in the lower urinary tract will remain the source of problems such as recurrent UTIs, requiring a secondary surgery at the bladder level to excise the ureterocele. For this reason, many are in favor of total reconstruction as a primary intervention, although it is long, complex, and often difficult in neonates and small infants. Ultimately, the approach should be individualized in each patient with the intent to accomplish all goals with the fewest possible interventions and complications.

Upper tract approach versus total primary reconstruction

For duplex system ectopic ureteroceles, the upper tract or simplified approach consists of a nephron sparing ureteropyelostomy/ureteroureterostomy or an upper pole hemi-nephroureterectomy (UHN) with near total ureterectomy. The ureteral stump is left open in the retroperitoneum (in the absence of VUR) and the ureterocele becomes a flaccid mucosal fold (Husmann et al., 1995; Jayram et al., 2011). The upper tract approach is more frequently considered in cases without associated VUR (or when it is mild) and depending on the UP renal function. When there is a reasonable function of the UP moiety, upper-to-lower ureteropyelostomy or ureteroureterostomy is the ideal choice. As mentioned earlier, more recently some authors have advocated this approach even in patients with minimal or no function of the UP moiety, leaving the non-functioning UP in place as the literature has shown no increased risk of malignancy, hypertension or infection (Mariyappa et al., 2014). It is important to note that this is done as a less aggressive intervention and not with the expectation of regaining renal function, as renal dysplasia is commonly present (Merlini & Chiesa, 2004). UHN is the traditional approach in patients with ectopic ureteroceles and non-functioning or minimally functioning associated UP moieties (contributing <10% of the total renal function) and without VUR in the lower pole moiety. In such patients, this approach has been shown to be curative 80%-85% of the time (Merlini & Chiesa, 2004). UHN is safe and complications are low and include retroperitoneal hematoma (3.5%), urinary leak (0%-12.5%), cysts at the resection margin of the remnant kidney needing regular imaging (0%-35%), loss of LP function due to vascular injury (6.8%), and recurrent UTI from persistent or new-onset VUR in other renal moieties or from stagnant urine within the ureteral stump and ureterocele (Mariyappa et al., 2014; Lopes et al., 2018; Jayram et al., 2011; Gundeti, Ransley, Duffy, Cuckow, & Wilcox, 2005), the latter being the most common indication for reoperation. The incidence of de novo VUR in the remaining lower pole moiety is reported to be around 15%-45% (Herz et al., 2016). Reoperation (15%-20%) involves bladder surgery in the form of excision of the ureterocele and associated distal ureteral stump and ipsilateral LP ureteral reimplantation.

In patients with duplex system ectopic ureteroceles and poorly functioning upper pole moieties where dilating reflux is present at diagnosis (whether in the LP moiety or contralateral ureter), UHN may still be considered although a secondary surgery will be necessary 84%-90% of the time (Merlini & Chiesa, 2004). Higher grades of reflux increase the chance of needing secondary interventions. The advantage of performing a simplified approach is that it may be the only intervention needed, particularly when VUR is absent. Even in those patients who need a secondary intervention, this approach may help delay future bladder surgery until the patient is older thus facilitating the technical aspects of surgery and decreasing the risk of bladder or sphincteric injury and subsequent incontinence (Husmann et al., 1995).

Total primary reconstruction is also an alternative in such patients particularly in cases where VUR is present, with a reported cure rate of 90% (De Jong, Dik, Klijn, Uiterwaal, & Gool, 2000; Decter, Sprunger & Holland, 2001). This combined upper and lower tract approach includes UHN or ureteropyelostomy/ureteroureterostomy depending on UP renal function, ureterocele excision, ipsilateral LP ureteral reimplantation (and contralateral when necessary) and in some cases reconstruction of the bladder neck. This approach is more technically demanding, involves longer operative times and poses a greater risk of complications such as urinary incontinence, especially in neonates and young infants (Gran, Kropp, Cheng, & Kropp, 2005). Proponents of this method argue that this surgery is safe and feasible nowadays since dedicated pediatric anesthesia is widely available and surgical specialists have greater expertise in neonatal reconstruction (Decter et al., 2001). Proponents of this approach argue that in most cases, patients undergo a single complete surgical intervention that addresses all issues associated with the ureterocele before age 1, obviating not only the need for further surgeries but also of repetitive invasive studies such as cystourethrograms and providing parents and patients with the unmeasurable psychological relief of "being cured" (Decter et al., 2001).

Upper pole hemi-nephrectomy

UHN for benign disease in children is most commonly indicated for poorly functioning duplex renal moieties in the setting of persistent hydronephrosis or UTI and due to urinary incontinence from an ectopic ureter (El-Ghoneimi et al., 2003; Singh et al., 2010). It can be accomplished in open fashion or with manual or robotic assisted laparoendoscopy. The surgical principles are similar irrespective of the approach. Frequently, a cystoscopy is performed to evaluate the ureteral orifice(s) insertion and the anatomy of the associated ureterocele. A retrograde pyelogram and ureteral stent placement of the LP ureter can be performed per surgeon's preference. Likewise, a retroperitoneal drain on the surgical bed is left at surgeon's discretion. Usually, a catheter for bladder drainage is left at least overnight.

Open upper pole hemi-nephrectomy

The patient is positioned supine with slight elevation of the ipsilateral side for a modified flank position. After access to the kidney is gained by surgeon's preference (usually a flank or anterior muscle-splitting approach) retroperitoneally, Gerota's fascia is opened to expose the kidney. The UP ureter is identified and dissected off the LP ureter, preserving carefully its blood supply (Fig. 8.1A). After dissecting the UP ureter as far down as possible, it is divided and dissected off the renal hilum. The ureter is then raised with light traction to help with further dissection of the UP renal vessels (Fig. 8.1B). The non-refluxing ureteral stump is left open to allow decompression of the ureterocele but the refluxing one is closed distally near the bladder wall. Often times a second, lower quadrant or suprapubic small incision is needed to remove a dilated, refluxing ureteral stump. The UP renal vessels are tied off, with care to preserve all LP vessels intact. When in doubt, a vessel can be

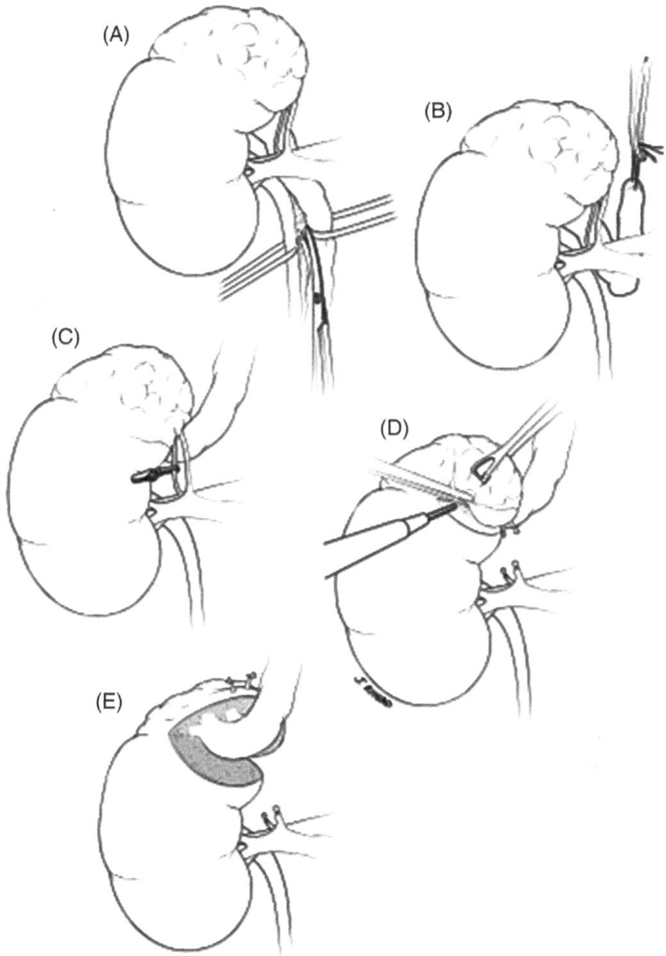

Figure 8.1 (A-E) Open upper pole hemi-nephrectomy. (A) The upper pole (UP) ureter is identified and dissected off the lower pole (LP) ureter, preserving carefully its blood supply. (B) The divided UP ureter is passed posteriorly to hilar vessels and raised with light traction to help further dissection of the UP renal vessels. (C) The UP renal vessels are tied off and divided, with care to leave any LP vessels intact. A bulldog clamp on the UP renal artery may be used to confirm parenchymal blanching prior to vessel division. (D) The ischemic UP parenchyma is separated along the indentation that demarcates it from the LP. (E) Upward traction on the UP ureter aids to visualize the almost bloodless plane of dissection between the UP renal pelvis and the LP parenchyma.

occluded with a bulldog clamp and parenchymal blanching can be observed to confirm its supply to the UP with subsequent ligation using absorbable suture (Fig. 8.1C). Then the indentation that demarcates the upper and lower poles are identified and the renal capsule is opened approximately 1 cm cranially from this location. Separation of the UP parenchyma is accomplished with a combination of finger dissection, careful sharp excision, and judicious use of electrocautery (Fig. 8.1D). Upward traction on the divided UP ureter allows visualization of the almost bloodless plane of dissection between the UP renal pelvis and the LP parenchyma (Fig. 8.1E). Care must be taken not to enter the collecting system of the LP, but if this is done, it can be oversown. If a lower pole external ureteral stent was placed at the time of cystoscopy, injection of methylene blue will confirm the integrity of the lower pole collecting system. The capsule may then be closed over if needed to achieve hemostasis.

Minimally invasive hemi-nephrectomy

Transperitoneal laparoscopy

Jordan and Winslow first described in 1993 the technique for transperitoneal (TP) laparoscopic hemi-nephrectomy in a pediatric patient using 10- and 12-mm ports and instruments (Jordan & Winslow, 1993). Since then, laparoscopic advances have allowed for using instruments as small as 3 mm (Breda, Lam, Veale, Lerman, & Schulam, 2007). In the transperitoneal approach, the patient is placed in a 30 to 40-degree lateral decubitus position with the ipsilateral side upward. The patient should be on a beanbag or similar padding and secured appropriately. The bed is tilted to a mild Trendelenburg position. Access and pneumoperitoneum is achieved per surgeon's preference, usually with insufflation at 10 mm Hg. Commonly, three ports are positioned to create a triangle with its base directed toward the ipsilateral renal pelvis (Fig. 8.2). In this manner, a 5 mm port is placed at the umbilicus (to accommodate the Harmonic scalpel). Next, a 3-5 mm port is placed in the midline equidistant between the umbilicus and xyphoid process, and another one in the ipsilateral lower quadrant along the mid clavicular line. The 5 mm, 0-degree laparoscope is used through the umbilical port, except for when the vessel sealer is used, at which time a 3-5 mm laparoscope is used through the alternative port. We prefer to use a bowel grasper and hook electrocautery for dissection, although dissection at the hilum is performed with the Maryland instead of the hook. To access the kidney, the peritoneum is incised along the white line of Toldt and the corresponding ascending or descending colon is freed and reflected medially. To gain access to the UP of the kidney, the liver or spleen is carefully retracted and when necessary, with careful dissection, the spleen is reflected cranially or the liver medially. An accessory 3 mm fanning liver retractor

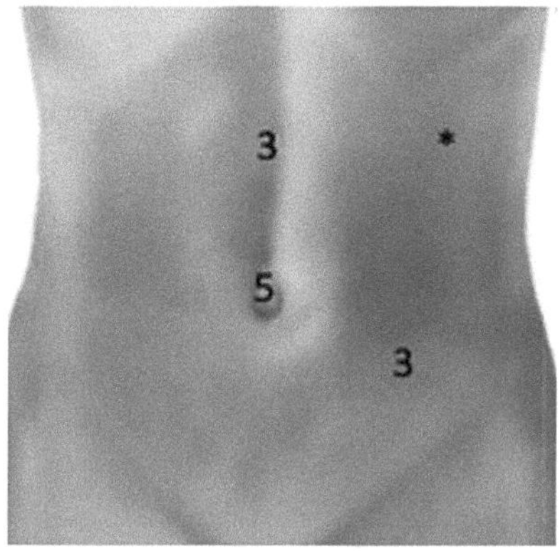

Figure 8.2 ***Port placement for laparoscopic transperitoneal upper pole hemi-nephrectomy.*** The ports are triangulated to point toward the ipsilateral renal pelvis: 5 mm at the umbilicus, 3 mm in the midline equidistant between the umbilicus and xyphoid process, and 3 mm in the ipsilateral lower quadrant along the mid clavicular line. Note that for older children, 5 mm ports and instruments are used.

may be used as needed. Gerota's fascia is opened and the UP is exposed. Attention is then turned to the duplicated ureters. The UP ureter is identified below the lower pole of the kidney and carefully dissected off from the LP ureter, preserving its blood supply. The gonadal vessels are encountered and spared. We often use a percutaneous suture, like 2-0 polydioxanone on a CT-1 needle, to pass as a sling under the UP ureter for proximal traction (Fig. 8.3A). After following the UP ureter to the renal hilum, the main renal vessels are identified. From these, the UP vessels are identified and isolated with careful dissection (Fig. 8.3B). The UP vessels are clipped and divided although some surgeons prefer to use the Harmonic scalpel. On occasion, the laparoscopic forceps may be used to occlude the UP renal artery and parenchymal blanching can be observed to confirm its supply to the UP prior to ligation. Dissection of the proximal UP ureter is continued from underneath the main renal vessels and on the plane of dissection between the UP renal pelvis and the renal parenchyma of the LP. The UP ureter is then transected and dissected as far distally as possible, close to the bladder wall. It is left open when non-refluxing or clipped when the ureterocele is refluxing. The proximal UP ureter is then passed under the main vessels. Applying upward traction of the divided UP ureter, the plane of dissection between the UP renal pelvis and the LP renal parenchyma is followed. This allows a safe, almost bloodless plane of dissection until the lateral aspect of the UP renal parenchyma can be divided with the Harmonic scalpel or electrocautery, away from the LP renal vessels. (Fig. 8.3C). Hemostasis is

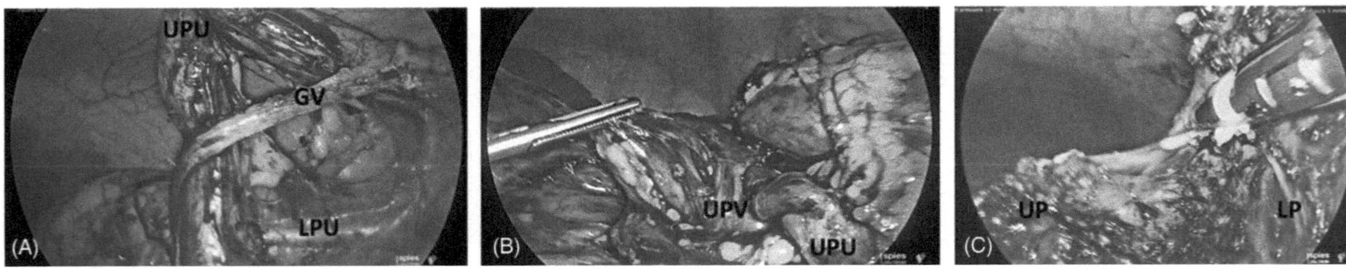

Figure 8.3 (A-C) Laparoscopic transperitoneal left upper pole hemi-nephrectomy. (A) The upper pole ureter (UPU) has been dissected from the lower pole ureter (LPU) and placed on traction with a 2-0 polydioxanone sling, noting that the gonadal vessels (GV) have been spared. (B) The upper pole vessels (UPV) have been dissected from the main renal vessels and are ready for clipping and division. (C) With the UP ureter on traction, the plane of dissection of the ischemic upper pole moiety (UP) is followed until it is completely excised from the lower pole moiety (LP).

achieved with judicious electrocautery. A topical hemostatic agent may be applied as needed. When the collecting system of the lower pole has been entered, it can be oversown with fine absorbable suture. The area of dissection is inspected after a few minutes have elapsed with the insufflation pressures decreased. The specimen is placed in a retrievable bag and removed through an expanded 5 mm umbilical port site. The rest of the ports are removed, and the skin and fascia are closed. A Foley catheter is left indwelling overnight. Unless the lower pole collecting system has been entered, we do not place a drain.

Retroperitoneal laparoscopy

In 2000, Miyasato et al. reported the first pediatric laparoscopic hemi-nephrectomy using the retroperitoneal (RP) approach (Miyazato et al., 2000). Although the RP approach most closely resembles that of open surgery, it is technically more demanding, especially due to a smaller working space.

For this technique, the patient is positioned in full flank position and the surgeon works facing the back of the patient. It is recommended to place the ports widely and to minimally protrude them so as to avoid instrument clash and maximize the retroperitoneal space (Goyal & Hennayake, 2010). First, a 1-2 cm incision is made at the level of the 12th rib, with a muscle-splitting fashion the retroperitoneum and Gerota's fascia are entered under direct vision, and a 5 mm port is placed. To prevent an air leak, a purse-string suture is placed around the port and the retroperitoneum is insufflated. Under direct vision, some blunt dissection with the camera port may be needed to develop the space. As a reference, we use the psoas muscle posteriorly and progressively create the retroperitoneal space with a combination of sweeping blunt dissection of Gerota's fascia off the muscle. Alternatively, the space may be developed using balloon dissection. Then, a 3-5 mm port is placed just lateral to the sacrospinalis muscle, and a third one about 1 finger-width medial and superior to the iliac crest (Fig. 8.4). The kidney is approached posteriorly, and the psoas muscle is used as a reference point. This approach allows direct visualization of the upper and lower pole ureters while avoiding dissection of the main renal vessels. In similar fashion as previously described, the UP vessels are identified, dissected, and clipped; the renal parenchyma is divided with the Harmonic scalpel and the UP ureter is dissected distally to a level proximal to the iliac vessels and divided. Of note, the UP moiety is freed totally from the peritoneum while it is left attached over the LP to minimize manipulation of the healthy moiety.

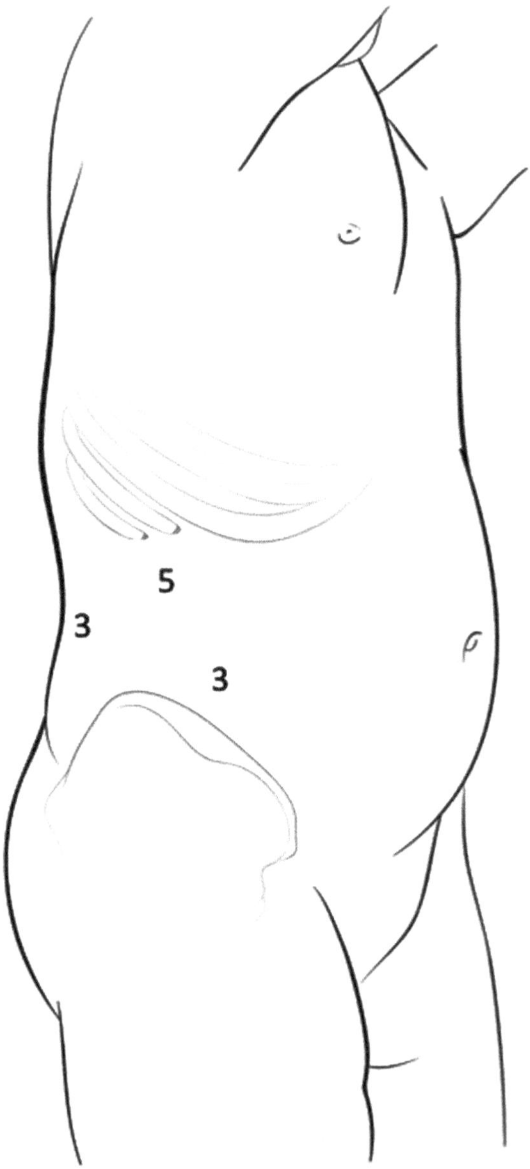

Figure 8.4 ***Port placement for laparoscopic retroperitoneal upper pole hemi-nephrectomy.*** A 5 mm port is placed at the level of the 12th rib. A 3 mm port is then placed just lateral to the sacrospinalis muscle. Finally, a 3 mm port is placed about a finger-width medial and superior to the iliac crest.

Transperitoneoscopic versus retroperitoneoscopic approaches

We have previously reported our experience with TP and RP laparoscopic approaches (Castellan et al., 2006). In experienced surgeon's hands, both approaches have been shown to be safe and feasible without significant differences in operative times, hospital stay or analgesic requirements (El-Ghoneimi et al., 2003; Castellan et al., 2006; Wallis et al., 2006). Advantages of the TP approach include a larger working space and easier visualization of the kidney and its renal hilum. However, a disadvantage is that in order to dissect the polar vessels and upper pole ureter, it requires dissection of the main renal vessels, risking vascular injury and the blood supply to the healthy lower pole parenchyma. In contrast, the RP approach, which more closely resembles the open technique, avoids potential injury to intraperitoneal organs and the development of bowel adhesions. It also allows rapid access to the renal hilum and dissection of the UP vessels is accomplished with little trauma to the main and other polar vessels. However, the RP approach is technically more demanding, the working space is smaller, it is more difficult to dissect the ureter all the way down to the bladder and there are more conversions to open or TP (mainly due to loss of working space secondary to development of pneumoperitoneum following a peritoneal tear) (Castellan et al., 2006; Wallis et al., 2006; Franc-Guimond, Al Hazmi, Gonzalez, Houle, & Barrieras, 2005). While several small series have shown the RP approach to be safe in infants without conversions nor significant complications (El-Ghoneimi et al., 2003; Lee, Retik, Borer, Diamond, & Peters, 2005), literature from the last decade and a half reports conversion rates around 6%-18% (Fernández Córdoba, Gonzálvez Piñera, & Argumosa Salazar, 2011; Esposito et al., 2015), with one study reporting as high as 21% (Leclair, Vidal, Suply, Podevin, & Héloury, 2009). Additionally, studies have shown increased morbidity in infants (Castellan et al., 2006; Wallis et al., 2006), including the possibility of damaging hemodynamic effects of retroperitoneal insufflation to the remaining LP moiety resulting in renal ischemia and total loss of function and new onset hypertension (Wallis et al., 2006). In 2015, Esposito et al reported their results of a multicenter international survey on 50 patients undergoing laparoscopic RP UHN performed by 6 experienced and well-known laparoscopic surgeons (Goyal & Hennayake, 2010). Their intraoperative complication rate was 14%, mainly peritoneotomies on children less than 12 months old. For this reason, we believe that experience in the laparoscopic technique of choice is imperative, but also that TP is indicated when complete excision of the ureter is required and with infants or small toddlers with large kidneys.

Robot-assisted laparoscopic surgery

Advantages of laparoscopy over the open technique in children include less blood loss, less postoperative analgesia and hospitalization times and improved cosmesis (El-Ghoneimi et al., 2003; Castellan et al., 2006; Casale & Kojima, 2009). Besides the significant experience needed to perform complex surgeries, drawbacks to laparoscopy include difficulty with intracorporal suturing and dissection. Using the robotic platform allows for improved ergonomics, magnified three-dimensional vision, and ease of instrument dexterity. By closely mimicking the open technique while maintaining the advantages of minimally invasive surgery, robot-assisted laparoscopic (RAL) surgery overcomes most of the issues associated with open or laparoscopic surgery and shortens the training curve. For RAL UHN, the patient is positioned in a modified flank fashion with careful padding and secured with tape (Fig. 8.5). Nowadays, RAL surgery in the pediatric population can be achieved with 5- or 8-mm instruments, depending on the daVinci surgical system used (Intuitive Surgical, Sunnyvale, CA, USA), although the Si platform does require a 12-mm camera port. Port placement also depends on the robotic platform used and surgeon's preference, but they are frequently placed in a configuration similar to other renal surgeries such as pyeloplasty or nephrectomy (Castellan et al., 2006). At the author's institution, the Xi platform is used, which employs 8-mm trocars and allows for linear midline port placement. For UHN, a camera port is placed at the umbilicus, two additional working ports (one sub-xyphoid and another equidistant between

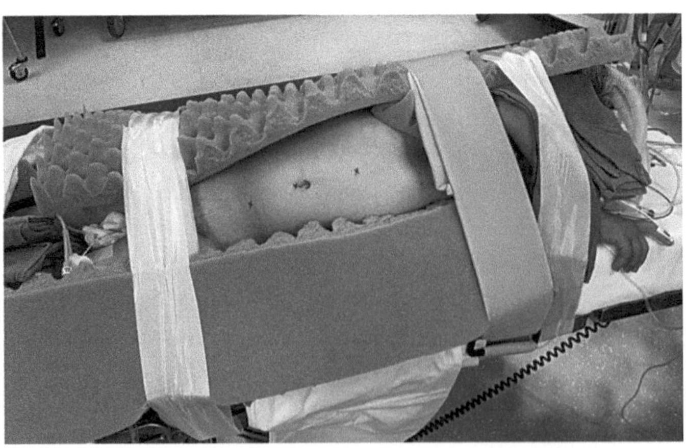

Figure 8.5 ***Positioning for robot-assisted laparoscopic upper pole hemi-nephrectomy.*** The patient is placed in the modified flank position, with the ipsilateral side upward, and carefully padded and secured. Anticipated port placement has been marked.

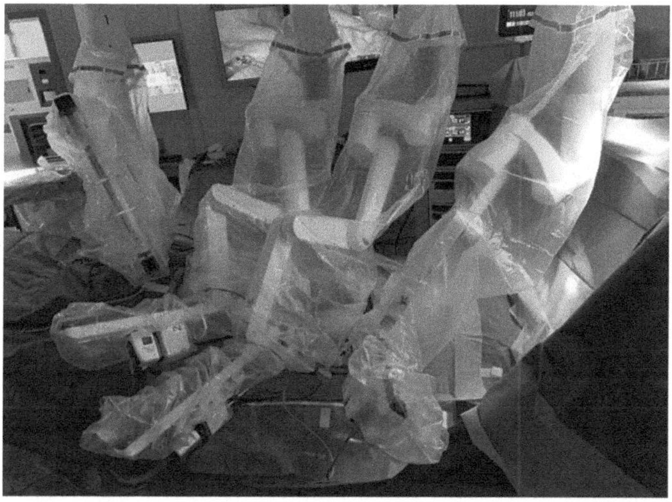

Figure 8.6 ***Docked robot configuration.*** **At the author's institution, regardless of laterality, the robotic cart is brought in from the left side. This is aided by deploying the boom as far as possible. The video tower stands on the right side, towards the head. The anesthesia team is at the patient's head.**

the umbilicus and bladder), and on occasion, a 3 or 5-mm bedside assistant port in the contralateral lower quadrant. CO_2 insufflation pressure is set in the 10–15 mmHg range. The robot device is frequently docked from the ipsilateral side, although different configurations may be employed as needed since the main boom can be rotated thus allowing the arms to reach at comfortable distances. At the author's institution, regardless of laterality, it is docked from the patient's left side due to the room configuration (Fig. 8.6). The surgical technique is similar as described laparoscopically (Fig. 8.7). The parenchymal transection can be accomplished with the monopolar hook or curved scissors (Casale & Kojima, 2009), or with the robotic vessel sealer. If desired, selective arterial mapping may be accomplished with the application of intraoperative near infrared fluorescence (NIRF) with indocyanine green (ICG) dye (Herz et al., 2016). This is also useful to assist in the dissection of the upper pole parenchyma. In 2004, Pedraza et al. reported the first case of bilateral RAL HN (Pedraza, Palmer, Moss, & Franco, 2004). Since then, studies have shown good outcomes with this modality, including using the retroperitoneal approach (Lee et al., 2009; Olsen & Jørgensen, 2005; Kutikov, Nguyen, Guzzo, Canter, & Casale, 2007; Kapoor & Elder, 2015). One criticism of using RAL surgery is its lengthy operative time (OT) as compared to open or even laparoscopic surgery (Lee et

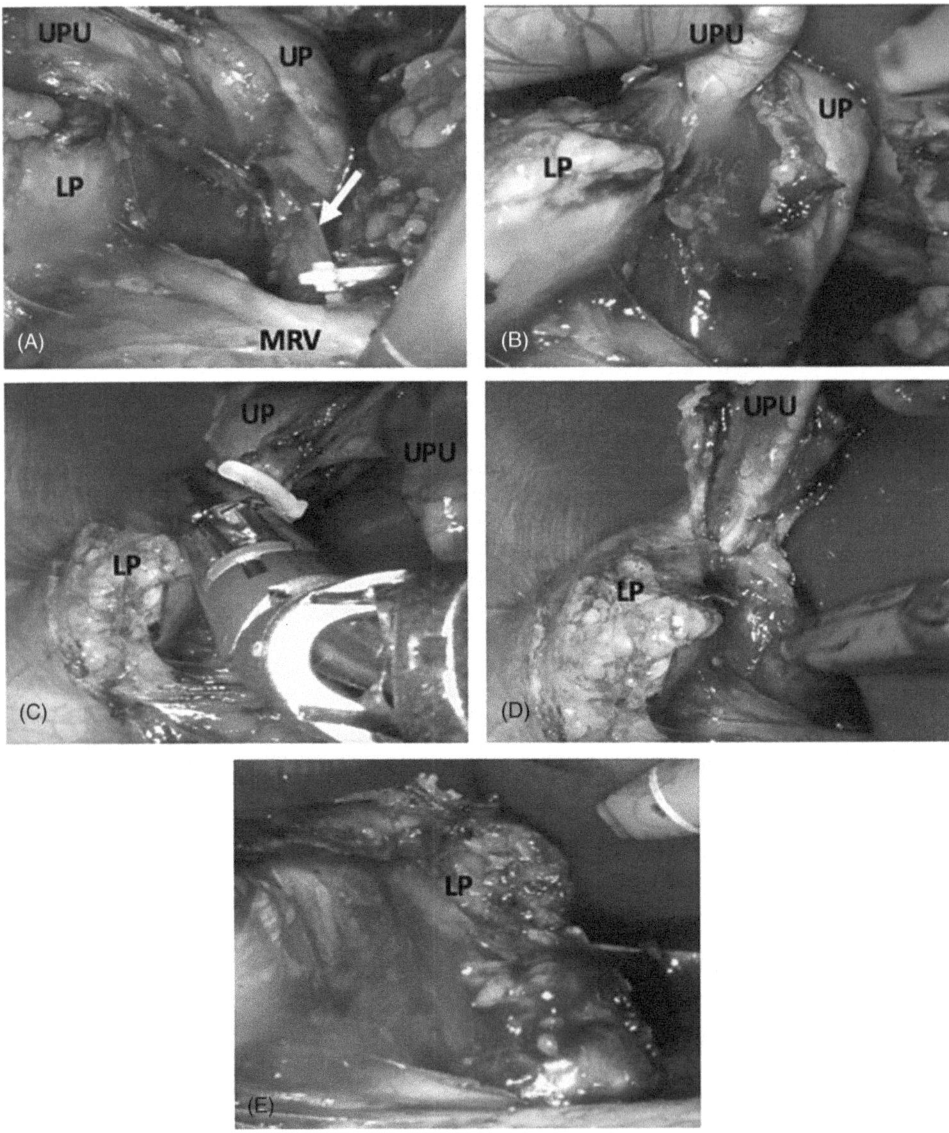

Figure 8.7 (A-E) Robotic transperitoneal right upper pole hemi-nephrectomy. (A) The divided upper pole ureter (UPU) has been dissected, divided, and passed posteriorly to the main renal vessels (MRV). The UP renal vein has been dissected, clipped, and divided. The exposed UP renal artery *(arrow)* will be clipped and divided. (B) With the UPU on light traction and gentle blunt dissection, the plane of dissection between the ischemic UP and the lower pole (LP) is visualized. (C) UP excision is accomplished with the robotic vessel sealer, following the almost bloodless plane between the UP renal pelvis and the LP parenchyma (D). (E) The remaining LP is inspected for hemostasis. If a lower pole external ureteral stent was placed at the time of cystoscopy, injection of methylene blue will confirm the integrity of the lower pole collecting system.

al., 2009). For example, one study reported a mean OT of 182 min using straight laparoscopy for UHN (García-Aparicio et al., 2010) while another reported a mean OT of 275 min for RAL HN (Lee et al., 2009). However, in the latter the authors reported an improvement in the OT as the learning curve for the procedure progressed, with the last few cases taking only 201 min. Most recently, Herz et al. performed an outcome comparison between RAL, open, and laparoscopic approaches to the management of children with duplex renal anomalies (Herz et al., 2016). In this series, 19 children underwent 20 RAL hemi-nephrectomies (17 UHN) with a median age of 3.1 years and a 95% surgical success. The mean operative time was 209 min, which shows the improving trend that comes with experience in the last few years of performing RAL HN.

Conclusions

Upper pole hemi-nephrectomy remains a viable option in the management of patients with ureteroceles subtending a non-functioning or poorly functioning duplex kidney. A minimally invasive approach is the gold standard for most children at this time. It can be performed safely in most situations where an open procedure is indicated. A transperitoneal or retroperitoneal approach to minimally invasive upper pole hemi-nephrectomy will yield the same results in expert hands. In patients without reflux into the upper pole ureter, the ureterocele is aspirated and the ureteral stump is left open in the retroperitoneum. In a subset of patients, a second lower tract procedure will be necessary to address issues of recurrent urinary tract infections and persistent dilating vesicoureteral reflux.

References

Breda, A., Lam, J. S., Veale, J., Lerman, S., & Schulam, P. G. (2007). Laparoscopic heminephrectomy for upper-pole moiety in children using a 3-mm laparoscope and instruments. *Journal of Endourology*, *21*(8), 883–885.

Casale, P., & Kojima, Y. (2009). Robotic-assisted laparoscopic surgery in pediatric urology: an update. *Scandinavian Journal of Surgery*, *98*(2), 110–119.

Castellan, M., Gosalbez, R., Carmack, A., Prieto, J., Perez-Brayfield, M., & Labbie, A. (2006). Transperitoneal and retroperitoneal laparoscopic heminephrectomy-what approach for which patient? *Journal of Urology*, *176*(6), 2636–2639.

Decter, R. M., Sprunger, J. K., & Holland, R. J. (2001). Can a single individualized procedure predictably resolve all the problematic aspects of the pediatric ureterocele? *The Journal of Urology*, *165*(6 Pt 2), 2308–2310.

De Jong, T. P., Dik, P., Klijn, A. J., Uiterwaal, C. S., & Gool, J. D. (2000). Ectopic ureterocele: results of open surgical therapy in 40 patients. *The Journal of Urology*, 2040–2043 discussion 3-4.

El-Ghoneimi, A., Farhat, W., Bolduc, S., Bagli, D., Mclorie, G., & Khoury, A. (2003). Retroperitoneal laparoscopic vs open partial nephroureterectomy in children. *BJU International, 91*(6), 532–535.
Esposito, C., Miyano, G., Caione, P., Escolino, M., Chiarenza, F., Riccipetitoni, G., et al. (2015). Retroperitoneoscopic heminephrectomy in duplex kidney in infants and children: results of a multicentric survey. *Journal of Laparoendoscopic & Advanced Surgical Techniques, 25*(10), 864–869.
Fernández Córdoba, M. S., Gonzálvez Piñera, J., Argumosa Salazar, Y., et al. (2011). Our experience in retroperitoneoscopic total or partial nephrectomies [in Spanish]. *Cirugía Pediátrica, 24*, 232–236.
Franc-Guimond, J., Al Hazmi, H., Gonzalez, R., Houle, A. M., & Barrieras, D. (2005). Laparoscopic partial nephrectomies and nephrectomies can be accomplished safely in the very young child. *Canadian Journal of Urology, 12*, 2787.
Gander, R., Asensio, M., Royo, G. F., & Lloret, J. (2016). Evaluation of the initial treatment of ureteroceles. *Urology, 89*, 113–117.
García-Aparicio, L., Krauel, L., Tarrado, X., Olivares, M., García-Nuñez, B., Lerena, J., et al. (2010). Heminephroureterectomy for duplex kidney: laparoscopy versus open surgery. *Journal of Pediatric Urology, 6*(2), 157–160.
Goyal, A., & Hennayake, S. (2010). Prone retroperitoneoscopic approach for heminephrectomy: specific advantages relating to access to vascular pedicle. *Journal of Pediatric Urology, 6*(2), 153–156.
Gran, C. D., Kropp, B. P., Cheng, E.Y., & Kropp, K. A. (2005). Primary lower urinary tract reconstruction for nonfunctioning renal moieties associated with obstructing ureteroceles. *Journal of Urology, 173*(1), 198–201.
Gundeti, M., Ransley, P., Duffy, P., Cuckow, P., & Wilcox, D. (2005). Renal outcome following heminephrectomy for duplex kidney. *Journal of Urology, 173*(5), 1743–1744.
Herz, D., Smith, J., Mcleod, D., Schober, M., Preece, J., & Merguerian, P. (2016). Robot-assisted laparoscopic management of duplex renal anomaly: comparison of surgical outcomes to traditional pure laparoscopic and open surgery. *Journal of Pediatric Urology, 12*(1), 44 e1-7.
Herz, D., Dajusta, D., Ching, C., & Mcleod, D. (2016). Segmental arterial mapping during pediatric robot-assisted laparoscopic heminephrectomy: a descriptive series. *Journal of Pediatric Urology, 12*(4), 266 e1-6.
Husmann, D. A., Ewalt, D. H., Glenski, W. J., & Bernier, P. A. (1995). Ureterocele associated with ureteral duplication and a nonfunctioning upper pole segment. *The Journal of Urology, 154*(2 Pt 2), 723–726.
Jayram, G., Roberts, J., Hernandez, A., Heloury, Y., Manoharan, S., Godbole, P., et al. (2011). Outcomes and fate of the remnant moiety following laparoscopic heminephrectomy for duplex kidney: a multicenter review. *Journal of Pediatric Urology*, 7(3), 272–275.
Jordan, G. H., & Winslow, B. H. (1993). Laparoendoscopic upper pole partial nephrectomy with ureterectomy. *Journal of Urology, 150*(3), 940–943.
Kapoor, V., & Elder, J. S. (2015). Simultaneous bilateral robotic-assisted laparoscopic procedures in children. *Journal of Robotic Surgery, 9*(4), 285–290.
Kutikov, A., Nguyen, M., Guzzo, T., Canter, D., & Casale, P. (2007). Laparoscopic and robotic complex upper-tract reconstruction in children with a duplex collecting system. *Journal of Endourology, 21*(6), 621–624.
Leclair, M., Vidal, I., Suply, E., Podevin, G., & Héloury, Y. (2009). Retroperitoneal laparoscopic heminephrectomy in duplex kidney in infants and children: a 15-year experience. *European Urology, 56*(2), 385–389.
Lee, R. S., Retik, A. B., Borer, J. G., Diamond, D. A., & Peters, C. A. (2005). Pediatric retroperitoneal laparoscopic partial nephrectomy: comparison with an age matched cohort of open surgery. *Journal of Urology, 174*(2), 702–712.

Lee, R. S., Sethi, A. S., Passerotti, C. C., Retik, A. B., Borer, J. G., Nguyen, H. T., & Peters, C. A. (2009). Robot assisted laparoscopic partial nephrectomy: a viable and safe option in children. *Journal of Urology, 181*(2), 823–829.

Lopes, R. I., Fernandez, N., Koyle, M. A., Santos, J. D., Perez, J., Romao, R., & Lorenzo, A. J. (2018). Clinical outcomes of the upper urinary tract after ureteral clipping for treatment of low functioning or nonfunctioning renal moieties. *Journal of Urology, 199*(2), 558–564.

Mariyappa, B., Barker, A., Samnakay, N., & Khosa, J. (2014). Management of duplex-system ureterocele. *Journal of Paediatrics and Child Health, 50*(2), 96–99.

Merlini, E., & Chiesa, P. L. (2004). Obstructive ureterocele-an ongoing challenge. *World Journal of Urology, 22*(2), 107–114.

Michaud, J. E., & Akhavan, A. (2017). Upper pole heminephrectomy versus lower pole ureteroureterostomy for ectopic upper pole ureters. *Current Urology Reports, 3*(18), 21.

Miyano, G., Takahashi, T., Nakamura, H., Doi, T., Okawada, M., Koga, H., et al. (2013). Retroperitoneoscopic nephrectomy/heminephrectomy in children planned, performed, and managed by supervised senior pediatric surgical trainees. *Journal of Laparoendoscopic & Advanced Surgical Techniques, 23*(8), 723–727.

Miyazato, M., Hatano, T., Miyazato, T., Kagawa, H., Yonou, H., & Ogawa, Y. (2000). Retroperitoneoscopic heminephrectomy of the right upper collecting system emptying into an ectopic ureterocele in a 5-year-old girl: a case report. *Hinyokika Kiyo, 46*, 413.

Olsen, L., & Jørgensen, T. (2005). Robotically assisted retroperitoneoscopic heminephrectomy in children: initial clinical results. *Journal of Pediatric Urology, 1*(2), 101–104.

Pedraza, R., Palmer, L., Moss, V., & Franco, I. (2004). Bilateral robotic assisted laparoscopic heminephroureterectomy. *Journal of Urology, 171*(6 Part 1), 2394–2395.

Sharifi, S. H., Nabavizadeh, B., Mozafarpour, S., & Kajbafzadeh, A. (2017). Laparoscopic selective clipping of upper moiety vasculature and ureter without partial nephrectomy: a novel technique for pediatric urinary incontinence due to ectopic ureter associated with poor functioning upper renal moiety. *Journal of Pediatric Urology, 13*(2), 217–218.

Singh, R. R., Wagener, S., & Chandran, H. (2010). Laparoscopic management and outcomes in non-functioning moieties of duplex kidneys in children. *Journal of Pediatric Urology, 6*(1), 66–69.

Wallis, M. C., Khoury, A. E., Lorenzo, A. J., Pippi-Salle, J. L., Bägli, D. J., & Farhat, W. A. (2006). Outcome analysis of retroperitoneal laparoscopic heminephrectomy in children. *Journal of Urology, 175*(6), 2277–2282.

CHAPTER 9

Ureteropyelostomy and ureteroureterostomy

Christopher S. Cooper[a], Angela M. Arlen[b]
[a]University of Iowa Hospitals and Clinics, Iowa City, IA, United States
[b]Department of Urology, Yale University School of Medicine, New Haven, CT, United States

Introduction

There are numerous options for managing ureteroceles associated with complete ureteral duplication. Embryologically, ureteral duplication is the result of abnormal branching of the ureteric bud off the mesonephric duct. The upper pole ureteric bud arises more proximal from the mesonephric duct relative to the lower pole ureteric bud leading to an upper pole orifice in a more caudal and medial position while the lower pole system is located more laterally, eponymously termed the Weigert-Meyer law. As a result, the upper pole is more commonly associated with ectopia or a ureterocele, while the lower pole tends to be associated with vesicoureteral reflux (VUR) due to shorter intramural tunnel length (Ambrose & Nicolson, 1964). Surgical decision-making for ureteroceles is influenced not only by a child's clinical course but also by the presence of reflux into any of the ureters and to a lesser extent by the function of the upper pole moiety, as well as surgeon and parental preference (Cohen, Juwono, Palazzi, Kaplan, & Chiang, 2015). Management goals include preservation of renal function, prevention of urinary tract infection (UTI) and obstruction, maintenance of continence, and minimizing surgical morbidity. If significant lower pole VUR is not a factor, ureteroureterostomy (UU) or ureteropyelostomy (UP) are valid reconstructive alternatives to ureterocele excision and reimplantation or ureterocele puncture.

Indications

Options for definitive surgical intervention of ureteroceles associated with duplicated system include various upper and lower ureteral approaches. An end-to-side ureteroureterostomy (UU) can be performed proximally or distally in appropriately selected duplex system. Based on surgeon

Ureteroceles
http://dx.doi.org/10.1016/B978-0-12-817888-1.00009-4

preference a ureteropyelostomy (UP) is also an option. Efficacy of ipsilateral ureteroureterostomy in cases of ureteral duplication is well-established, and several recent studies have reported similar success with this technique for the management of ureteroceles (Kawal et al., 2019; Lee et al., 2015; Timberlake & Corbett, 2015) UU or UP can be considered when obstruction of the upper pole moiety is the only feature encountered with a ureterocele in a duplex system, regardless of upper pole function and ureteral size disparity (McLeod, Alpert, Ural, & Jayanthi, 2014; Sheth et al., 2019). The contribution of the affected upper pole moiety to the global renal function typically ranges between 4% and 8% (Merlini & Lelli Chiesa, 2004). Therefore preservation of this relatively small amount of upper pole function in patients with ureteroceles is not a major determinant of operative management. In the past, upper pole function was strongly considered when determining if an upper pole heminephrectomy should be considered, however, more recent studies demonstrate that a poorly functioning upper pole that is draining well is not associated with significant risks. Kawal and colleagues reported that upper pole function of less than 10% (range 0%–4%, median 0%) had no impact on outcomes in children undergoing UU for various indications (Kawal et al., 2019). Therefore, avoidance of ablation procedures irrespective of upper moiety function is now favored by most pediatric urologists.

Advantages of UU/UP include a single stage definitive procedure, and the ability to remain extravesical, thus avoiding the potential morbidity inherent with a transvesical approach or upper pole ablative procedures (Gran, Kropp, Cheng, & Kropp, 2005; Lashley, McAleer, & Kaplan, 2001; Michaud & Akhavan, 2017). UU offers the theoretical advantage of preserving any functioning upper pole parenchyma, while also avoiding renal surgery with potential risk of renal bleeding and damage to the lower pole moiety. UP, or a more proximally performed UU, may provide the benefit of reduced urinary stasis within the dilated upper pole ureter (Baek, Au, Huang, & Koh, 2017). UU and UP can be performed via an open, laparoscopic or robot-assisted approach.

General surgical principles for ureteroureterostomy and ureteropyelostomy

- Cystoscopy with stent or ureteral catheter placement to distinguish upper from lower pole ureters can be considered, particularly in cases where ureteral size discrepancy is lacking

- Dissection kept to a minimum with limited direct handling of ureters
- Mobilization should be focused on the upper pole ureter so as to avoid disruption of the lower pole with care taken to preserve ureteral adventitia
- Upper pole ureter is often markedly dilated, so the recipient lower pole ureter or renal pelvis must be adequately incised to allow for optimal end-to-side anastomosis
- Anastomosis may be performed in a running or interrupted fashion; the authors prefer a small absorbable monofilament suture in running fashion with the knots placed extraluminally
- After the posterior anastomosis is complete, a double J stent is placed in the lower pole system if not previously done, and the anterior anastomosis performed; the authors leave an indwelling stent for 4–6 weeks postoperatively
- Anastomosis should be tension free and water tight
- Placement of a closed suction or penrose drain is at the surgeon's discretion; the authors do not routinely leave a drain
- Renal-bladder ultrasound is obtained approximately 4–6 weeks after stent removal, with additional imaging obtained as clinically indicated

Surgical approach

Many factors are considered when determining the optimal surgical approach for a given child. Compared to open surgery, minimally invasive approaches offer several advantages including smaller incisions and more rapid convalescence. However, this must be balanced with increased operative times and cost, particularly when a robotic approach is employed.

Open

Distal UU (Fig. 9.1) can be performed retroperitoneally through a Pfannenstiel, Gibson, or small inguinal herniorrhaphy type incision (Prieto, Ziada, Baker, & Snodgrass, 2009). UP is typically performed through a flank incision. Success rates of >90% have been reported in children undergoing open UU for ureterocele (Lashley et al., 2001; Lee et al., 2015; Prieto et al., 2009). In many series, patients undergoing open surgery are younger than those undergoing a minimally invasive approach, and as the majority of ureteroceles are diagnosed prenatally, open UU remains common (Le & Chiang, 2018; Lee et al., 2015).

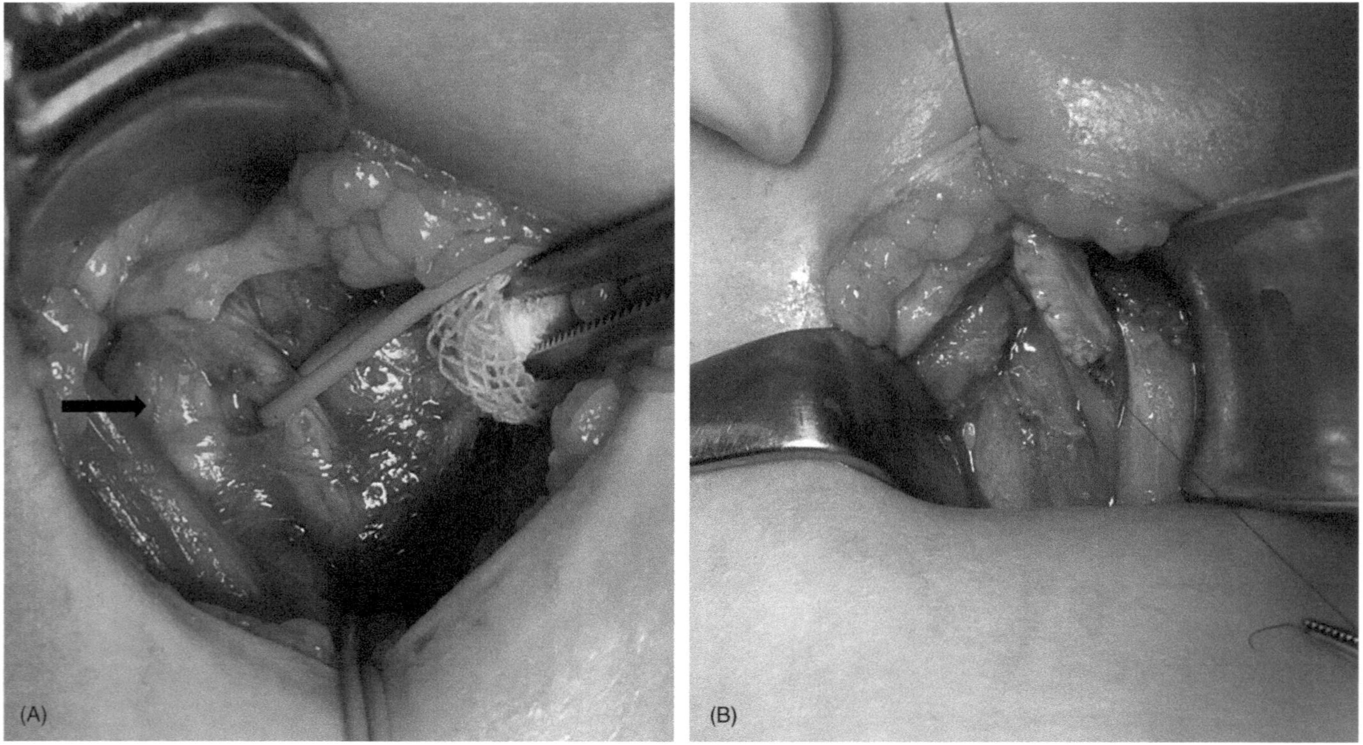

Figure 9.1 Open distal ureteroureterostomy. Upper vessel loop encircling a normal caliber lower pole ureter; the dilated upper pole ureter is denoted with an arrow (A). The upper pole ureter is anastomosed to the lower pole in a running, water-tight fashion using monofilament suture (B).

Laparoscopic or robotic

Minimally invasive surgical techniques are increasingly utilized in the pediatric population for complex reconstruction, and pure laparoscopic UU has been described (González & Piaggio, 2007; Olguner et al., 2012). However, given the need for delicate intracorporeal suturing, robotic surgery is particularly advantageous for these reconstructive procedures (Bilgutay & Kirsch, 2019; Van Batavia & Casale, 2014). In comparison to open surgery, robotic surgery has been associated with decreased morbidity, less postoperative pain, lower analgesic requirements, quicker postoperative recovery, and shorter hospital stays. While improved cosmesis is often considered a benefit over open surgery, the location in addition to scar size, impacts patient satisfaction (Garcia-Roig et al., 2017). It is also hypothesized that robotic UU/UP offers more complete visualization of the ureter throughout its course (Kawal et al., 2019). The technique for robotic UU/UP can be performed in fashion similar to pyeloplasty. Alternatively, UU can also be performed more distally (Fig. 9.2), with port placement similar to that of extravesical ureteral reimplantation (Bilgutay & Kirsch, 2019). It is imperative to correctly identify each ureter, which can easily be facilitated by cystoscopic placement of a temporary catheter in the recipient lower pole ureter prior positioning for port placement.

Outcomes/Complications

Lashley and colleagues reported a 92% success rate in children undergoing open UU for ureterocele, and similar success rates have been published regardless of open or minimally invasive surgical approach (Kawal et al., 2019; Lashley et al., 2001; Lee et al., 2015; McLeod et al., 2014). Clinical and radiographic outcomes have also been shown to be independent of renal function of the obstructed moiety (Kawal et al., 2019; McLeod et al., 2014).

Duplex systems treated with UU or UP are left with a distal ureteral stump, which may develop pathology such as urinary tract infections or even stones requiring additional surgery. There is also the possibility of new onset VUR, ureteral anastomotic stricture (potentially damaging the lower pole), the so-called "yo-yo reflux," and urine leak (Michaud & Akhavan, 2017). Concerns that a retained poorly functioning or nonfunctioning upper pole moiety will lead to hypertension, proteinuria, or malignancy have not been substantiated (Chacko, Koyle, Mingin, & Furness, 2007; Husmann, 1998; Lashley et al., 2001).

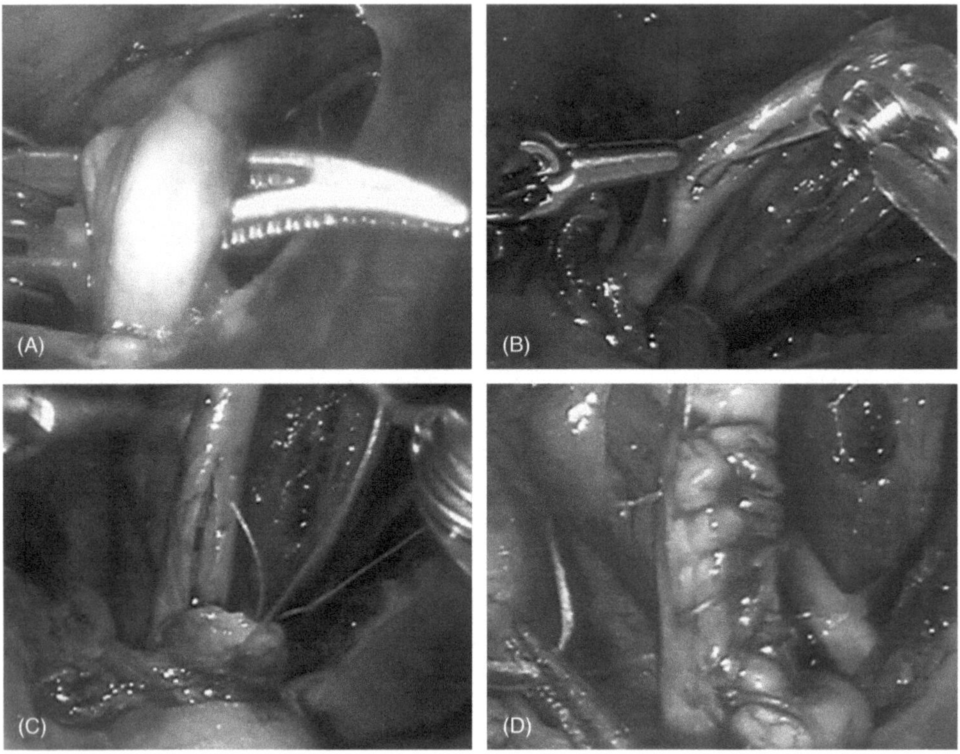

Figure 9.2 Robotic distal ureteroureterostomy. Both ureters are identified in the pelvis (A). After transection of the upper pole ureter, the receipt lower pole ureter (with temporary stent in place) is incised (B). The upper pole ureter is then anastomosed to the lower pole in a running, water-tight fashion (C, D).

Early complications

- Occur within the first few days following surgical intervention, are typically transient, and are often stent related
- Anastomotic leaks have been reported in up to 13% of cases
- Postoperative UTI can often be avoided by obtaining a preoperative urine culture

Late complications

- Ureteral stricture (~2%) is the most serious late surgical complication of UU or UP; complete obstruction necessitates drainage and the majority of these patients will require reoperation
- Theoretical development of de novo ureteroureteral "yo-yo" VUR, presenting as increased lower pole hydroureteronephrosis or UTI from urinary stasis, after UU may require medical or surgical treatment

- No documented increased risk of UTI or stones in the absence of persistent obstruction despite in situ ureterocele

Conclusions

An individualized approach based on clinical course, associated anomalies such as VUR, should guide operative management in children with ureteroceles. Ureteroureterostomy and ureteropyelostomy are safe and effective options in those patients with an obstructive upper pole in the absence of lower pole reflux, irrespective of upper pole function.

References

Ambrose, S. S., & Nicolson, W. P. (1964). Ureteral reflux in duplicated ureters. *Journal of Urology*, *92*(5), 439–443 https://doi.org/10.1016/S0022-5347(17)63983-1.

Baek, M., Au, J., Huang, G. O., & Koh, C. J. (2017). Robot-assisted laparoscopic pyeloureterostomy in infants with duplex systems and upper pole hydronephrosis: variations in double-J ureteral stenting techniques. *Journal of Pediatric Urology*, *13*(2), 219–220 https://doi.org/10.1016/j.jpurol.2016.12.008.

Bilgutay, A. N., & Kirsch, A. J. (2019). Robotic ureteral reconstruction in the pediatric population. *Frontiers in Pediatrics*, 7, 85 https://doi.org/10.3389/fped.2019.00085.

Chacko, J. K., Koyle, M. A., Mingin, G. C., & Furness, P. D. (2007). Ipsilateral ureteroureterostomy in the surgical management of the severely dilated ureter in ureteral duplication. *The Journal of Urology*, *178*(4 Pt. 2), 1689–1692 https://doi.org/10.1016/j.juro.2007.05.098.

Cohen, S. A., Juwono, T., Palazzi, K. L., Kaplan, G. W., & Chiang, G. (2015). Examining trends in the treatment of ureterocele yields no definitive solution. *Journal of Pediatric Urology*, *11*(1), 29.e1–29.e6 https://doi.org/10.1016/j.jpurol.2014.09.001.

Garcia-Roig, M. L., Travers, C., McCracken, C., Cerwinka, W., Kirsch, J. M., & Kirsch, A. J. (2017). Surgical scar location preference for pediatric kidney and pelvic surgery: a crowdsourced survey. *The Journal of Urology*, *197*(3 Pt. 2), 911–919 https://doi.org/10.1016/j.juro.2016.11.033.

González, R., & Piaggio, L. (2007). Initial experience with laparoscopic ipsilateral ureteroureterostomy in infants and children for duplication anomalies of the urinary tract. *Journal of Urology*, *177*(6), 2315–2318 https://doi.org/10.1016/j.juro.2007.01.177.

Gran, C. D., Kropp, B. P., Cheng, E. Y., & Kropp, K. A. (2005). Primary lower urinary tract reconstruction for nonfunctioning renal moieties associated with obstructing ureteroceles. *The Journal of Urology*, *173*(1), 198–201 https://doi.org/10.1097/01.ju.0000148374.64478.b5.

Husmann, D. A. (1998). Renal dysplasia: the risks and consequences of leaving dysplastic tissue in situ. *Urology*, *52*(4), 533–536 https://doi.org/10.1016/S0090-4295(98)00289-1.

Kawal, T., Srinivasan, A. K., Talwar, R., Chu, D. I., Long, C., Weiss, D., …, & Shukla, A. R. (2019). Ipsilateral ureteroureterostomy: does function of the obstructed moiety matter? *Journal of Pediatric Urology*, *15*(1), 50.e1–50.e6 https://doi.org/10.1016/j.jpurol.2018.08.012.

Lashley, D. B., McAleer, I. M., & Kaplan, G. W. (2001). Ipilateral ureteroureterostomy for the treatment of vesicoureteral reflux or obstruction associated with complete ureteral duplication. *Journal of Urology*, *165*(2), 552–554 https://doi.org/10.1097/00005392-200102000-00067.

Le, H. -K., & Chiang, G. (2018). Long-term management of ureterocele in duplex collecting systems: reconstruction implications. *Current Urology Reports*, *19*(2), 14 https://doi.org/10.1007/s11934-018-0758-3.

Lee, N. G., Corbett, S. T., Cobb, K., Bailey, G. C., Burns, A. S., & Peters, C. A. (2015). Bi-institutional comparison of robot-assisted laparoscopic versus open ureteroureterostomy in the pediatric population. *Journal of Endourology*, *29*(11), 1237–1241 https://doi.org/10.1089/end.2015.0223.

McLeod, D. J., Alpert, S. A., Ural, Z., & Jayanthi, V. R. (2014). Ureteroureterostomy irrespective of ureteral size or upper pole function: a single center experience. *Journal of Pediatric Urology*, *10*(4), 616–619 https://doi.org/10.1016/j.jpurol.2014.05.003.

Merlini, E., & Lelli Chiesa, P. (2004). Obstructive ureterocele-an ongoing challenge. *World Journal of Urology*, *22*(2), 107–114 https://doi.org/10.1007/s00345-004-0407-y.

Michaud, J. E., & Akhavan, A. (2017). Upper pole heminephrectomy versus lower pole ureteroureterostomy for ectopic upper pole ureters. *Current Urology Reports*, *18*(3), 21 https://doi.org/10.1007/s11934-017-0664-0.

Olguner, M., Akgür, F. M., Türkmen, M. A., Siyve, S., Hakgüder, G., & Ates¸, O. (2012). Laparoscopic ureteroureterostomy in children with a duplex collecting system plus obstructed ureteral ectopia. *Journal of Pediatric Surgery*, *47*(4), e27–e30 https://doi.org/10.1016/j.jpedsurg.2011.12.010.

Prieto, J., Ziada, A., Baker, L., & Snodgrass, W. (2009). Ureteroureterostomy via inguinal incision for ectopic ureters and ureteroceles without ipsilateral lower pole reflux. *The Journal of Urology*, *181*(4), 1844–1848 https://doi.org/10.1016/j.juro.2008.12.004.

Sheth, K. R., White, J. T., Janzen, N., Koh, C. J., Austin, P. F., Roth, D. R., …, & Seth, A. (2019). Evaluating outcomes of differential surgical management of nonfunctioning upper pole renal moieties in duplex collecting systems. *Urology*, *123*, 198–203 https://doi.org/10.1016/j.urology.2018.06.028.

Timberlake, M. D., & Corbett, S. T. (2015). Minimally invasive techniques for management of the ureterocele and ectopic ureter: upper tract versus lower tract approach. *The Urologic Clinics of North America*, *42*(1), 61–76 https://doi.org/10.1016/j.ucl.2014.09.006.

Van Batavia, J. P., & Casale, P. (2014). Robotic surgery in pediatric urology. *Current Urology Reports*, *15*(5), 402 https://doi.org/10.1007/s11934-014-0402-9.

CHAPTER 10

Bladder level surgery for ureteroceles

Yvonne Y. Chan, Earl Y. Cheng
Division of Pediatric Urology, Ann and Robert H. Lurie Children's Hospital of Chicago, Chicago, IL, United States

Introduction

Historical perspective

The role of lower urinary tract reconstruction in ureterocele management has evolved over the last 50 years. Prior to the advent of transurethral techniques, ureteroceles were treated with open excision or incision depending on the surgeon's preference. Gross and Clatworthy advocated for open bladder incision rather than excision of ureteroceles, fearing the latter would result in the development of vesicoureteral reflux (VUR). However, Snyder and Johnston favored complete ureterocele excision with concurrent ureteral reimplantation (Gross & Clatworthy, 1950; Snyder & Johnston, 1978). Hendren and Mitchell, in turn, advocated for complete reconstruction, which involves heminephrectomy of the nonfunctional moiety, ureteral reimplantation, and complete ureterocele excision to reduce risks of postoperative infections, bladder outlet obstruction, and bladder diverticula (Hendren & Mitchell, 1979).

With the development and popularization of the transurethral incision approach in the 1990s, surgeons began to favor endoscopic intervention as the primary approach for decompression of the obstructed upper pole moiety (Pohl, 2011; Rich, Keating, Snyder, & Duckett, 1990; Shimada, Matsumoto, & Matsui, 2007). Lower urinary tract reconstruction, which can be technically challenging and carries the risk of injury to the urinary sphincter and vagina, became reserved for cases of persistent VUR or obstruction following endoscopic and/or upper tract intervention (Castagnetti, Cimador, Sergio, & de Grazia, 2004; Cooper et al., 2000; de Jong, Dik, Klijn, Uiterwaal, & van Gool, 2000; Decter, Sprunger, & Holland, 2001; Rich et al., 1990; Shekarriz, Upadhyay, Fleming, Gonzalez, & Barthold, 1999).

In the past, combined upper and lower tract interventions were often performed as a single-stage or two-stage procedures for fear that leaving a

Ureteroceles
http://dx.doi.org/10.1016/B978-0-12-817888-1.00010-0

non-functional renal moiety would increase risk of urinary tract infection, hypertension, or proteinuria (Beganovic, Klijn, Dik, & De Jong, 2007; Decter, Roth, & Gonzales, 1989; Gran, Kropp, Cheng, & Kropp, 2005; Hendren & Mitchell, 1979; Shekarriz et al., 1999). This combined approach results in two incisions and sometimes two anesthetics for each patient. Recent evidence has suggested that removal of the non-functional upper pole moiety may not be necessary for all patients. Gran et al. evaluated the outcomes of 16 patients with duplex system ureteroceles and associated non-functional renal moieties who underwent lower urinary tract reconstruction without upper tract intervention. Complete ureterocele excision, bladder neck reconstruction, and ureteral reimplant as single stage procedures were performed. Upper tract dilatation improved in all patients, and none experienced loss of renal function or required additional interventions (Gran et al., 2005). There were no complications related to leaving the non-functional upper pole moiety in situ. Thus, observation is now favored as other studies also continue to demonstrate that leaving non-functional renal moieties in situ after successful endoscopic decompression is safe (Chertin et al., 2005).

The evolution of lower urinary tract reconstruction's role in ureterocele management continues even until today. Historically, VUR, whether ipsilateral or contralateral, has been a strong indication for lower urinary tract reconstruction, but this indication has become controversial in light of recent evolutions in VUR management (Weiss, 2015). For typical duplex system ureteroceles, the upper pole is commonly associated with the ureterocele and is obstructed. Pre-intervention reflux is present in the ipsilateral lower pole in 50% of cases, and contralateral reflux is seen in 25% of cases. New onset reflux may occur in the ipsilateral lower pole, in the previously obstructed upper pole, and in the contralateral ureter(s) after endoscopic intervention (Shokeir & Nijman, 2002). Many studies are now showing spontaneous resolution or improvement of VUR after endoscopic decompression. Jesus et al. retrospectively evaluated 60 children with duplex system ureteroceles at a single institution who underwent endoscopic intervention. Forty patients (67%) had pre-intervention ipsilateral lower pole or contralateral VUR, of whom 15% improved and 42.5% remained stable after intervention. New onset VUR developed into the ipsilateral upper moiety in 3 patients and into a contralateral renal unit in 4 patients with pre-intervention VUR. New onset VUR developed in 17 patients without pre-intervention VUR, but the authors did not specify the specific renal units associated with these occurrences. Overall, 24 patients had new onset VUR into a new renal unit, and spontaneous resolution or improvement was seen in 16/24 (67%) of the group (Jesus et al., 2011). Similarly, Shimada et al. noted that only 10% of patients who underwent transurethral ureteral incision had persistent reflux post-intervention (Shimada et al., 2007). Adorisio et al. also noted spontaneous resolution of ipsilateral, lower moiety VUR in 71% of patients with pre-intervention VUR who underwent endoscopic intervention. VUR developed in the ipsilateral ureterocele moiety after intervention in 5 patients, which spontaneously resolved in two patients by 6 months follow-up (Adorisio et al., 2011). VUR was not found to be a predictor of breakthrough febrile urinary tract infections (UTI) in a cohort of 102 patients with antenatally detected ureteroceles managed with active surveillance for a median of 1.2 years (Andrioli et al., 2018). In light of these findings, the role of VUR as an indication for lower urinary tract reconstruction is no longer absolute and is being re-evaluated.

Indications for lower urinary tract reconstruction

Over the last 50 years, lower urinary tract reconstruction has evolved from being the primary intervention for all ureteroceles to the second line treatment for cases of persistent obstruction or reflux. Its role has continued to evolve as emerging studies challenge the need for concomitant upper tract ablation or intervention for all VUR. Furthermore, open or robotic ureteroureterostomy (as discussed in another chapter) is now playing a role in the primary management of ureteroceles. As such, fewer lower urinary tract reconstructions are being performed. Nevertheless, it still plays a role in the management of ureteroceles with residual high-grade reflux, symptomatic reflux with recurrent urinary tract infections (UTI), or residual ureteral obstruction following primary intervention. This chapter will review the surgical approaches for lower urinary tract reconstruction in ureterocele management and the outcomes of each approach. As lower urinary tract reconstructions are rarely performed for single system ureteroceles, the following discussions will focus on surgical management of duplex system ureteroceles.

Intravesical approaches to surgical intervention

Lower urinary tract reconstruction for ureterocele management involves ureterocelectomy, possible bladder neck reconstruction, and ureteral reimplantation with or without tapering. Bladder neck reconstruction is typically indicated in cases of ectopic ureteroceles with extension of the ureterocele through the bladder neck into the urethra or in cases of large intravesical ureteroceles that cause bladder neck distortion. In general, reconstruction is performed after 12 months of age due to concerns of bladder size in infants. Historical perspectives have raised concerns for bladder injury with reconstruction in small bladders and risk of bladder distortion and possible denervation following reconstruction surgery. With refined techniques, the challenges of early reconstruction have been traversed, and de Jong et al. demonstrated that reconstruction in neonates is feasible (de Jong et al., 2000). Nevertheless, intervening prior to one year of age should be reserved for cases mandated by the clinical scenario for which less invasive alternatives are not feasible or effective.

Ureterocele excision

Ureterocele excision may be accomplished via complete excision or marsupialization. Complete excision involves removal of all aspects of the ureterocele wall, leaving a defect in the posterior bladder wall that requires reconstruction. Surgeons that favor complete excision as opposed to marsupialization feel that reconstruction of the posterior bladder wall may help prevent bladder diverticula and bladder dysfunction with detrusor reapproximation (Lewis et al., 2008). In cases of ectopic ureteroceles that extend down the urethra, complete excision can be more challenging. Extra care must be taken when dissecting the posterior ureterocele wall off the bladder especially near the bladder neck as this area abuts the anterior vaginal wall in females and rectum in males. Perforation may increase the risk of fistula formation. Due to the technical challenges of complete excision and associated risks, marsupialization has been entertained as an alternative. In 1969, Johnston and Johnson proposed the technique of ureterocele "unroofing" or marsupialization. This technique does not require the extensive dissection necessary for complete ureterocele excision and avoids the need for bladder base reconstruction (Johnston & Johnson, 1969). Care must be taken during marsupialization to not leave a loose lip of ureterocele wall, which may cause bladder outlet obstruction.

Complete ureterocele excision surgical approach

Complete excision is typically performed via an intravesical approach. The bladder is exposed through a transverse lower abdominal incision. Cystotomy is performed with a sagittal incision extending down to but not beyond the bladder neck. The intravesical portion of the ureterocele is then visualized. As mentioned previously, the upper pole is typically associated with the ureterocele and is obstructed (Fig. 10.1A). The anterior

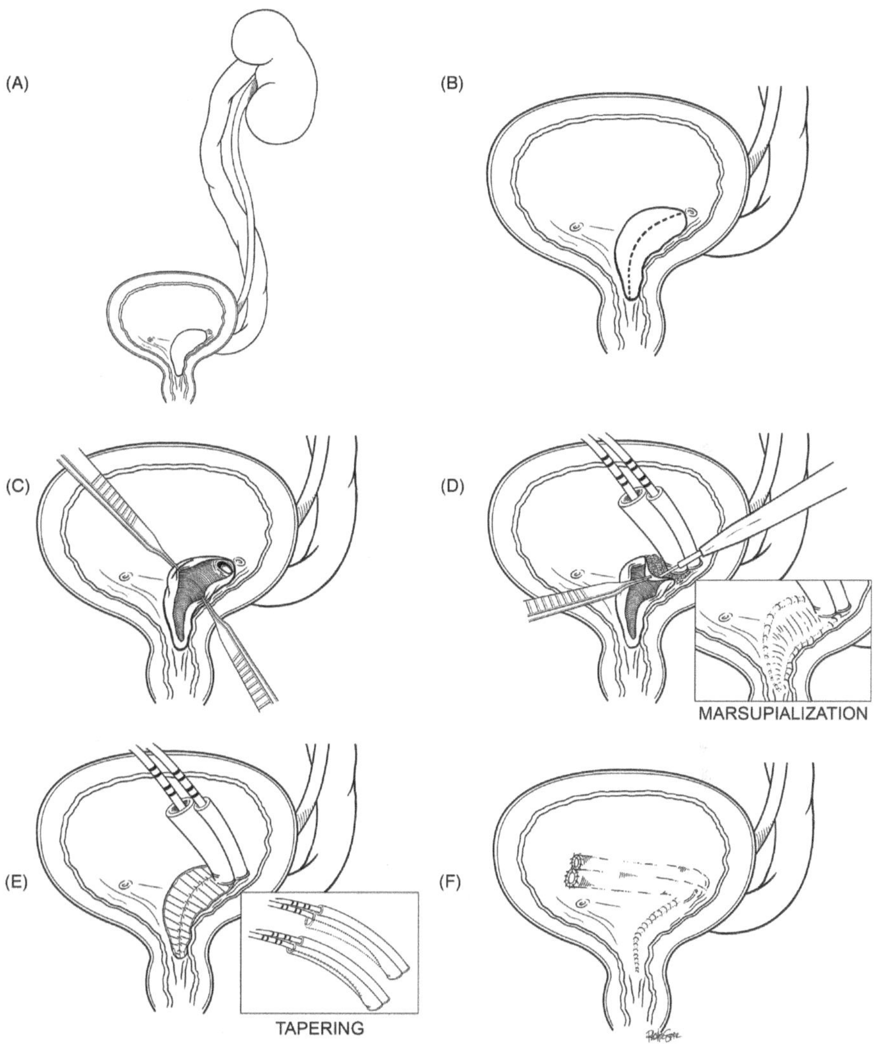

Figure 10.1 ***Open ureterocele repair.*** (A) For a typical duplex system ureterocele, the upper pole is associated with the ureterocele and is obstructed. (B) After the intravesical portion of the ureterocele is exposed, the anterior wall of the ureterocele is incised. (C) The upper pole ureteral orifice is visualized, and both ureteral orifices are intubated with feeding tubes/stents. (D) The duplex ureters are dissected through the ureteral hiatus as a single unit. The uretorecele is then dissected off the bladder in the plane between the posterior ureterocele wall and the bladder wall. Marsupialization insert: If marsupialization is performed, only the anterior and lateral ureterocele walls are excised. The posterior ureterocele wall is left intact and the mucosal edges are sutured down to the bladder. (E) After the ureterocele is excised, the detrusor defect is closed is closed in layers. Ureteral tapering insert: Ureteral tapering may be performed by excising excess ureteral tissue opposite to the abutting ureter. The ureter can be tapered over and 8-10 Fr stent. (F) Cross-trigonal common sheath reimplant is performed.

wall of the ureterocele is then incised (Fig. 10.1B). Both ureteral orifices are intubated with feeding tubes/stents (Fig. 10.1C). The ureters are then dissected through the ureteral hiatus. In general, separation of the duplex ureters is not recommended since it may compromise the common blood supply. Once the ureters are freed, the ureterocele may be dissected off the bladder in a cephalad to caudal manner. Dissection is performed along a plane between the posterior ureterocele wall and the bladder wall sharply or with electrocautery with a fine tip Bovie (Fig. 10.1D). To prevent inadvertent injury to the vagina in girls or rectum in boys when dissecting with the Bovie, dissection should be carried closer to the ureterocele wall. In cases of ectopic ureteroceles, the ureterocele may extend further down into the urethra. Should the ureterocele be small, one may retract the ureterocele cephalad for complete excision. When determining the extent of distal dissection, the risk of sphincter injury with resulting incontinence must be considered. Should the ureterocele extend significantly beyond the bladder neck, complete excision to the level of the bladder neck followed by marsupialization of the distal remaining ureterocele is recommended.

It is important to not leave a lip of ureterocele mucosa as this may result in bladder outlet obstruction via a valve mechanism. After excision, one may pass a catheter antegrade through the bladder neck into the urethra to evaluate for any obstruction. Transurethral endoscopic evaluation may also be performed after reconstruction, and transurethral resection of any leaflets of residual tissue may be pursued if indicated.

Once the ureterocele is excised, the detrusor defect is closed in layers with 4-0 or 5-0 absorbable sutures in an interrupted fashion (Fig. 10.1E). The mucosal defect in the trigone is then closed with 5-0 absorbable sutures.

Marsupialization surgical approach

Marsupialization does not require the extensive dissection necessary for complete ureterocele excision. The bladder is exposed and cystotomy performed as described above. The anterior and lateral walls of the ureterocele are excised off the bladder and urethral walls while leaving the posterior ureterocele wall intact. The ureterocele mucosal edges are then sutured down to the bladder with 5-0 or 6-0 absorbable sutures to prevent the creation of obstructing folds and to ensure hemostasis (Fig. 10.1D insert). Care must also be taken to not leave a lip of ureterocele wall, which can cause obstruction. Evaluation for residual tissue may be performed as described earlier.

Bladder neck reconstruction

In cases of complete ureterocele excision for ectopic ureteroceles extending beyond the bladder neck, bladder neck reconstruction is necessary. This involves re-approximation of the bladder neck with 4-0 or 5-0 absorbable sutures. Care must be taken to realign the sphincter fibers. In some cases, the bladder neck is completely distorted by the ectopic ureterocele, which may cause an element of sphincter insufficiency. In such cases, a meticulous, layered reconstruction is necessary to restore the continence mechanism.

Reimplantation and ureteral tapering

Ureteral reimplantation is performed following ureterocele excision/marsupialization, repair of the detrusor defect, and bladder neck reconstruction (as needed). Common sheath re-implantation is usually performed in a cross-trigonal manner (Fig. 10.1F). This approach is preferred as it avoids reimplantation in the area of the ureterocele, which often has a weakened muscular wall. The ureteral neo-orifice is created superior and lateral to the contralateral orifice. The submucosal tunnel is then developed and the ureters are brought through the tunnel. The neo-orifice is matured with interrupted 5-0 absorbable sutures. Care must be taken to ensure the ureters are straight and not kinked. The authors also prefer to place a single stitch securing the ureters to the new muscular hiatus. Reimplantation of the contralateral ureter is performed if reflux is seen on pre-operative imaging. In the event that cross-trigonal reimplantation cannot be performed, reimplant in the location of the ureterocele after excision and bladder reconstruction would be another option. If this is not possible, reimplant may be performed cephalad to the trigone in more healthy tissue, similar to that seen in the Young-Dees-Leadbetter bladder reconstruction.

If ureteral dilation is present, tapering of the upper and/or lower pole ureters is necessary. Dissection along the distal periureteral sheath directly opposite to the abutting ureter should be performed sharply to reveal the dilated ureter. The ureter and the surrounding adventitial tissue are then opened in the same plane to preserve the blood supply. Care must be taken to not disrupt where the ureters are attached as this contains the common ureteral blood supply. Excess ureteral tissue may then be excised and the ureter tapered over an 8-10 Fr stent (Fig. 10.1E insert). The proximal two-thirds of the tapering can be performed with running interlocking 5-0 absorbable suture in two layers. The first layer involves full-thickness bites of ureter followed by a second layer of Lembert sutures involving the adventitia to cover the suture line. The distal one-third should be completed with interrupted sutures to allow for ureteral trimming which may be necessary during reimplant. Reimplantation may then be performed as described above. Stents are left in place post-operatively.

Some authors have also discussed plication of the ureter as opposed to removal of excess ureteral tissue. However, the bulky, plicated ureter may be difficult to reimplant in a thickened bladder, which may occur in cases in which the ureterocele has resulted in bladder outlet obstruction. Yamazaki et al. favored the Paquin reimplant technique with plication (Yamazaki, Setoguchi, Yago, Suzuki, & Toma, 1999). With the Paquin reimplantation technique, the ureter is mobilized extravesically, and an incision is made in the posterior aspect of the bladder to facilitate the creation of a long tunnel. A nipple is formed on the distal ureteral end using a wedge of the distal ureter to further prevent reflux (Dewan, 2000).

Closure and follow-up

After completion of the reconstruction, the bladder is then closed in three layers. If bladder neck reconstruction is not performed, a Foley catheter and Penrose drain are left overnight. Should bladder neck reconstruction be performed, the Foley and Penrose are left for 3–5 days. If externalized stents are placed, they are left in for 7–14 days. Should a double J ureteral stent be placed, this is removed endoscopically 6 weeks following repair.

Renal and bladder ultrasound is obtained 1 month post-operatively to evaluate for residual obstruction. It is important to note that if high-grade reflux was present pre-operatively, hydronephrosis may be present on ultrasound. The hydronephrosis should not be worse than the overall degree of dilation seen with reflux on cystogram. Worsening post-operative hydronephrosis may suggest obstruction. Should there be concern for residual symptomatic reflux, a voiding cystourethrogram (VCUG) may be obtained 3–6 months post-operatively after adequate bladder healing. The decision to obtain a VCUG must be weighed against the risks of radiation, keeping in mind that the presence of low-grade reflux will not affect management in most cases. As continence may be affected by reconstruction, patients are typically followed until they are toilet trained. Should continence be achieved and imaging reveal no major concerns at that time, the patient does not require long-term urological follow-up unless subsequent issues arise.

Extravesical approach to surgical intervention

Some surgeons have adopted a modified extravesical approach to lower urinary tract reconstruction. For cases in which there is adequate upper pole function and no reflux in the ipsilateral lower pole, Merguerian et al. reported their experience with extravesical upper pole dismembered re-implant. For this procedure, the upper pole ureter is carefully dissected off the lower pole ureter in an extravesical fashion to preserve the blood supply to both ureters. The upper pole ureter is dissected distal to the lower pole ureter and transected. Ureteral tapering is then performed if necessary. The ureterocele is left intact. The detrusor muscle is then incised longitudinally 4–5 times the ureteral diameter on the lateral wall. The upper pole ureter is advanced 1 cm and re-implanted into the bladder. The detrusor muscle is then closed over the ureter. For cases in which lower pole reflux is present, they favor upper pole dismembered re-implant and lower pole non-dismembered re-implant (Merguerian et al., 2003).

Outcomes and complications

Review of the literature reveals that outcomes are often reported through retrospective reviews of single institution case series. The heterogeneity of ureterocele characteristics, surgeon preference in management, and outcomes reported make comparisons between various surgical approaches challenging. Complications following lower urinary tract reconstruction may include persistent obstruction, residual reflux, urinary tract dysfunction, voiding dysfunction, or incontinence. The outcomes of various series are summarized in Table 10.1.

Complete ureterocele excision

Overall, lower urinary tract reconstruction with complete ureterocele excision is a successful procedure. It is typically pursued as the definitive treatment for patients who have failed other forms of primary intervention. Cooper et al. noted that bladder level surgeries (re-implant and bladder neck reconstruction) accounted for 36% of secondary surgeries in their series of 44 patients (Cooper et al., 2000). Wang et al. followed their cohort of 30 patients and noted that reconstruction with ureterocele excision and lower pole reimplant was performed as the primary intervention in only one patient. However, lower urinary tract reconstruction was ultimately

Table 10.1 Outcomes after lower urinary tract reconstruction for ureterocele management.

Study	Patients	Surgical approach	Outcomes
Complete ureterocele excision			
Decter et al. (1989)	N = 20 No previous intervention.	N = 12 (60%) with complete ureterocele excision, reimplant, and upper tract intervention N = 2 (10%) with common sheath reimplant only N = 6 (30%) with complete ureterocele excision and reimplant without upper tract intervention	**Reoperation rate**: 10% **Postoperative VUR:** -N = 2 (10%) VUR transient postoperative reflux -N = 1 (5%) with persistent VUR requiring repeat reimplant **Obstruction:** -N = 1 (5%) with atrophy of ipsilateral lower pole **Voiding dysfunction/incontinence:** -N = 1 (5%) with urinary retention from incomplete ureterocele excision requiring endoscopic excision and repeat reimplant of contralateral ureter for obstruction **Complications:** Not reported
Shekarriz et al. (1999)	N = 26 No previous intervention.	All had complete excision, reimplant with or without upper tract surgery N = 10 (38%) underwent concomittent upper tract surgery	**Reoperation rate:** 4% **Postoperative VUR:** -N = 2 (8%) with persistent post-operative VUR requiring reoperation **Obstruction:** Not reported **Voiding dysfunction/incontinence:** -All continent -N = 1 (4%) with post-operative UTI -N = 1 (4%) with bladder instability **Other complications:** Not reported

de Jong et al. (2000)	N = 40 23% with prior TUI/ nephrostomy tube drainage. 3% with prior vesicostomy and ureteral drainage.	N = 39 (98%) underwent complete excision with ureteral reimplant N = 5 (12%) patients with lower-tract reconstruction and UP procedures as staged procedures N = 16 (41%) with colposuspension performed when no vesicourethral angle seen after bladder base reconstruction	**Reoperation rate:** 33% **Postoperative VUR:** -N = 9 (22.5%) with persistent reflux requiring reoperation **Obstruction:** -N = 1 (2%) with loss of renal function **Voiding dysfunction/incontinence:** -N = 4 (10%) with bladder outlet obstruction requiring reoperation -No incontinence -N = 16 (38%) with postoperative UTI -N = 5 (13%) with voiding dysfunction requiring biofeedback **Other complications:** -N = 1 (2%) with diverticulum
Decter et al. (2001)	N = 10 No prior intervention	N = 9 (90%) underwent complete excision, reimplant of ipsilateral LP with or without upper tract surgery N = 1 (10%) underwent complete excision and reimplant	**Reoperation rate:** 10% **Postoperative VUR:** None **Obstruction:** Not reported **Voiding dysfunction/incontinence:** -N = 1 (10%) required marsupialization of incompletely excised distal end of cecoureterocele -None with issues with toilet training **Other complications:** Not reported

(*Continued*)

Table 10.1 Outcomes after lower urinary tract reconstruction for ureterocele management. (*Cont.*)

Study	Patients	Surgical approach	Outcomes
Gomes et al. (2002)	N = 18 No prior intervention.	All had complete excision with bladder neck reconstruction. Reimplant in 45% systems; heminephrectomy in 55% systems	**Reoperation rate:** 6% **Postoperative VUR:** -100% of those with preoperative VUR resolved -N = 1 (6%) with new onset VUR post-operatively which is managed with prophylaxis **Obstruction:** -N = 1 (6%) with bilateral ureteral obstruction requiring reimplant **Voiding dysfunction/incontinence:** -N = 1 (6%) with dysfunctional voiding and required CIC **Other complications:** Not reported
Upadhyay et al. (2002)	N = 14 No prior intervention.	All common sheath reimplant with or without ureterocele excision and bladder neck reconstruction N = 7 (50%) with upper tract intervention	**Reoperation rate:** 7% **Postoperative VUR:** N = 1 (7%) with postoperative new onset VUR **Obstruction:** Not reported **Voiding dysfunction/incontinence:**-N = 1 (7%) with incontinence (diagnosed with vesicovaginal fistula) -N = 2 (21%) with voiding dysfunction -N = 3 (27%) with postoperative UTI **Other complications:** -N = 1 (7%) with vesicovaginal fistula

Gran et al. (2005)	N = 16 N = 13 (81%) treated with prior TUI.	All underwent complete excision, bladder neck reconstruction, reimplant. N = 14 (88%) required UP ureteral tapering.	**Reoperation rate:** 0% **Postoperative VUR:** -None with new onset reflux -N = 1 (6%) with persistent reflux **Obstruction:** None **Voiding dysfunction/incontinence:** -N = 2 (13%) with postoperative UTI **Other complications:** Not reported
Merguerian et al. (2005)	N = 30 cases N = 10 (33%) with prior TUI, N = 7 (23%) with prior partial nephrectomy, N = 1 (3%) with prior cutaneous ureterostomy.	N = 15 (50%) with extravesical reconstruction w/o ureterocele excision N = 15 (50%) with complete excision and intravesical reimplant	For extravesical reconstruction: **Reoperation rate:** None **Postoperative VUR:** -N = 1 (7%) postoperative low grade VUR **Obstruction:** None **Voiding dysfunction/incontinence:** -No incontinence -No postoperative UTI **Other complications:** Not reported For complete excision: **Reoperation rate:** 20% Post-operative VUR: None **Obstruction:** N = 2 (13%) with complete obstruction and requiring complete nephrectomy **Voiding dysfunction/incontinence** N = 3 (20%) incontinence with resolution in N = 2 (13%). One was awaiting reconstruction for incontinence management **Other complications:** N = 1 (7%) of iatrogenic ureteral perforation

(*Continued*)

Table 10.1 Outcomes after lower urinary tract reconstruction for ureterocele management. (*Cont.*)

Study	Patients	Surgical approach	Outcomes
Shimada et al. (2007)	N = 59 Unspecified portion of patients underwent prior TUI.	All underwent complete excision. Reimplant in N = 55 (93%). Nephroureterectomy of the UP performed in N= 4 (7%).	**Reoperation rate:** None **Postoperative VUR:** None **Obstruction:** None **Voiding dysfunction/incontinence:** -N = 4 (7%) with incontinence -N = 8 (14%) with voiding dysfunction -N = 12 (20%) with post-operative UTI **Other complications:** Not reported
Beganovic et al. (2007)	N = 52 N = 2 (4%) underwent prior UP resection	Total upper and lower urinary tract reconstruction in N = 50 (96%). N = 2 (4%) staged procedures.	**Reoperation rate:** 38% **Postoperative VUR:** N = 7 (13%) with postoperative reflux (N = 1 new onset) **Obstruction:** N = 1 (2%) with loss of renal function **Voiding dysfunction/incontinence:** -N = 3 (6%) incontinence -N = 7 (13%) with bladder outlet obstruction requiring additional procedures -N = 9 (17%) postoperative UTI **Other complications:** N = 1 (2%) with diverticulum
Mariyapp et al. (2013)	N = 10 Preoperative intervention not specified.	Complete excision and common sheath reimplant in N = 5 (50%) Complete excision, UP nephroureterectomy, LP reimplant in N = 5 (50%)	**Reoperation rate:** None **Postoperative VUR:** Not reported **Obstruction:** Not reported **Voiding dysfunction/incontinence** No postoperative UTI **Other complications:** None

Lee et al. (2015)	N = 39 N = 24 (62%) with prior TUI	Complete excision and common sheath reimplant	**Reoperation rate:** 18% **Postoperative VUR:** N = 2 (5%) requiring endoscopic treatment **Obstruction:** N = 3 (8%); 2 required repeat reimplant. 1 underwent nephrectomy **Voiding dysfunction/incontinence** -N = 2 (5%) with bladder outlet obstruction requiring endoscopic excision of ureterocele remnant -N = 3 (7.5%) for daytime incontinence -N = 9 (23%) postoperative UTI with N = 5 (12%) postoperative febrile UTI **Other complications:** Not reported
Marsupialization			
Scherz et al. (1989)	N = 34 No prior intervention	Combined upper tract and lower tract intervention (ureterocele marsupialization in all lower tract surgeries)	**Reoperation rate:** 12% **Postoperative VUR:** N = 4 (12%) requiring intervention **Obstruction:** Not reported **Voiding dysfunction/incontinence:** -No incontinence **Other complications:** None

(Continued)

Table 10.1 Outcomes after lower urinary tract reconstruction for ureterocele management. (*Cont.*)

Study	Patients	Surgical approach	Outcomes
Lewis et al. (2008)	N = 57 N = 27 (82%) in excision group with prior TUI N = 12 (50%) in marsupialization group with prior TUI	N = 33 (58%) with complete excision N = 24 (42%) with marsupialization Re-implantation performed in all patients	Complete excision **Reoperation rate:** Not reported **Postoperative VUR:** N = 4/30 (13%) with persistent reflux **Obstruction:** N = 9/30 (30%) with worsening hydronephrosis **Voiding dysfunction/incontinence:** -None incontinent -N = 6/25 (24%) with voiding dysfunction -N = 13/30 with recurrent UTI **Other complications:** N = 1/30 (3%) with diverticulum <u>Marsupialization</u> **Reoperation rate:** Not reported Postoperative VUR: N = 3/21 (14%) with persistent reflux Obstruction: N = 2/23 (9%) with worsening hydronephrosis **Voiding dysfunction/incontinence:** -N = 1/21 (5%) incontinent -N = 5/20 (25%) with voiding dysfunction -N = 11/24 (46%) with recurrent UTI **Other complications:** N = 2/21 (9%) with diverticulum

Cohen et al. (2015)*	N = 47 No previous interventions.	Marsupialization N = 35 (77.8%) Ureterectomy N = 8 (17.8%) TUI/Incision N = 1 (2.2%) Other N = 1 (2.2%)	**Reoperation rate:** 21.3% **Postoperative VUR:** -N = 8 (17.7%) (N = 2 (4.7%) contralateral, N = 6 (13%) ipsilateral) **Obstruction:** Not reported **Voiding dysfunction/incontinence:** Not reported **Other complications:** Not reported

CIC, clean intermittent catheterization; *LP*, lower pole; *TUI*, transurethral incision; *UTI*, urinary tract infection; *UP*, upper pole; *VUR*, vesicoureteral reflux implemented in 14 patients as the definitive treatment, although two patients still remained on antibiotic prophylaxis following surgery. Seventy-two percent of the cohort who achieved "treatment free" status, defined as no longer requiring urological oversight, achieved this after undergoing ureterocele excision and lower pole ureteral reimplant (Wang, Greenfield, Williot, & Rutkowski, 2008).

Success of reconstruction was often defined by reoperation rates, which ranged from 0% to 38% (Beganovic et al., 2007; de Jong et al., 2000; Gomes, Mendes, Castro, & Reis, 2002; Shekarriz et al., 1999). In the two studies with the highest reoperation rates, over 70% of the procedures were endoscopic interventions for VUR, which may not necessarily prompt intervention with recent changes to the management of VUR (Beganovic et al., 2007; de Jong et al., 2000). Lee et al. reported a more recent series of 39 patients with ectopic, duplex system ureteroceles who underwent complete excision and common sheath reimplant without upper tract intervention. They noted improvement in upper tract dilatation in 92% of patients. The authors reported an 18% reoperation rate of which 28% of the cases were for VUR (Lee et al., 2015).

Upper tract obstruction following lower urinary tract reconstruction is relatively rare. Most series report incidence of 0–10% (Lee et al., 2015; Shimada et al., 2007). Lewis et al. did note that 30% of those in their series who underwent complete excision had worsened hydronephrosis on follow-up renal ultrasound. However, the authors did not discuss need for intervention and long-term renal function in this group (Lewis et al., 2008). Despite the relatively low rates, follow up ultrasound for worsening hydronephrosis after reconstruction is imperative. Endoscopic intervention or repeat open surgery may be necessary for persistent obstruction. Febrile urinary tract infections may also suggest residual upper tract obstruction and/or reflux and mandates radiographic evaluation and surgical intervention as warranted.

Absence or resolution of VUR is also a common outcome evaluated following lower urinary tract reconstruction. However, characteristics of the VUR including reflux grade and the renal unit with reflux are not often reported. This may be secondary to the limitations of retrospective case series. Rates of post-operative VUR following lower urinary tract reconstruction range from 0% to 22% (de Jong et al., 2000; Decter et al., 2001). As mentioned previously, it was often an indication for reoperation. With the evolvement of VUR management, low grade and asymptomatic reflux will likely be managed conservatively.

Historically, there were concerns that bladder level surgeries for ureteroceles in infants conferred higher morbidity with risk of vaginal injury or disruption of the trigone leading to bladder dysfunction. Urinary incontinence and bladder dysfunction were not outcomes measured in many series. In those that evaluated these outcomes, lasting incontinence rates ranged from 0% to 8%. Voiding dysfunction rates ranged from 0% to 14% (Beganovic et al., 2007; de Jong et al., 2000; Decter et al., 2001; Lewis et al., 2008; Merguerian et al., 2003; Shekarriz et al., 1999; Shimada et al., 2007) (Table 10.1). It is important to note that patients with bilateral ectopic ureteroceles may be at significant increased risk for bladder and sphincter dysfunction. Sherman et al. reported the outcomes of twelve patients with bilateral ectopic ureteroceles. Although primary therapy varied, the group ultimately all underwent complete ureterocele excision with bilateral ureteral reimplant and upper tract intervention. Voiding dysfunction and incomplete emptying was seen in 70% of patients. None had urinary incontinence (Sherman, Stock, & Hanna, 2003). The presence of voiding dysfunction in 6% of patients with ectopic ureteroceles regardless of surgical intervention have led some authors to wonder whether bladder dysfunction is an inherent factor of the disease rather than the by-product of surgery (Holmes, Coplen, Strand, Husmann, & Baskin, 2002). Further studies are necessary to better elucidate whether this hypothesis is true. Among all the studies, only one case of vesicovaginal fistula and two cases of bladder diverticulum were reported following primary ureterocele excision (Beganovic et al., 2007; Lewis et al., 2008; Upadhyay et al., 2002).

It is, however, important to note that there are unique cases of ectopic ureteroceles that cause complete disruption of the bladder neck leading to intractable incontinence from sphincter insufficiency. Husmann et al. reported that 6% of their patients who underwent partial nephroureterectomy without lower urinary tract intervention for ureterocele management developed incontinence. Endoscopic evaluation revealed poor coaptation of the bladder neck. All patients required bladder neck reconstruction with excision of the weakened bladder neck musculature. Four underwent concomitant anterior bladder neck suspension and two required artificial urinary sphincters (Husmann, Ewalt, Glenski, & Bernier, 1995). These patients with structurally poor bladder neck competence, albeit rare, need to be managed carefully. Although not specifically described for ureteroceles, the need for bladder neck closure with Mitrofanoff diversion has been described for cases of intractable incontinence due to ectopic ureters and may be considered in extreme cases (Jayanthi, Churchill, McLorie, & Khoury, 1995).

Marsupialization

Marsupialization has been reported to be effective and safe with minimal risk of vaginal or sphincter injury. Scherz et al. reported a success rate of 86% with no incidences of vesicovaginal fistula or urethrocutaneous fistula (Scherz, Kaplan, Packer, & Brock, 1989). To date, there have been few studies that have evaluated the outcomes of ureterocelectomy via marsupialization (Table 10.1). Lewis et al. compared the results of marsupialization with complete ureterocele excision. The authors did not note any significant differences in rates of VUR, bladder diverticulum, continence, UTI, or voiding dysfunction. However, improved hydronephrosis was higher in the marsupialization group compared to complete excision (91% vs. 70%; $P = 0.047$). As the surgical approach had no effect on the incidence of bladder dysfunction, which is fairly high in this series, the authors suspected that impaired bladder health may be secondary to the poor development of the muscular layer on the posterior ureterocele wall and is an issue inherent in the disease (Lewis et al., 2008). Cohen et al. also reported their outcomes of 120 patients treated for ureteroceles of whom 36 were treated with marsupialization. The authors classified the procedures into categories of upper tract approach, lower tract approach, combined upper and lower tract approach, and staged lower tract approach. Based on this classification, the outcomes of those who only underwent marsupialization could not be definitively ascertained. However, the lower tract approach group was found to have a 21.3% rate of reoperation, although the nature of these procedures was not specified (Cohen, Juwono, Palazzi, Kaplan, & Chiang, 2015).

Extravesical reimplant

Merguerian et al. reported success in their series of patient who underwent extravesical dismembered ureteral reimplant without ureterocele excision. They noted postoperative low grade VUR in 7% of patients who did not require additional intervention. There were otherwise no incidences of incontinence, complications, or post-operative UTI (Merguerian et al., 2003).

Conclusions

Over the last 50 years, management of ureteroceles has become more individualized and conservative. With refinement of endoscopic techniques, lower urinary tract reconstruction is no longer a common primary intervention for ureteroceles. The evolution of VUR management may further limit its need as a secondary intervention in the future. Nevertheless, multiple case series have shown that lower urinary tract reconstruction, whether by complete excision or marsupialization, remains an effective and definitive option for ureterocele management. Despite the complexity of the procedure, complications such as vesicovaginal fistulas and bladder diverticula remain rare. This may be reflective of the fact that surgeons tend to prefer and practice procedures they are most comfortable with, and these series reflect the results of seasoned practitioners. Nevertheless, understanding the surgical principles of lower urinary tract reconstruction remains important, as there will always be patients who would benefit from this surgery.

Acknowledgment

We would like to acknowledge our artist Rachel Stork who illustrated the images of this chapter.

References

Adorisio, O., Elia, A., Landi, L., Taverna, M., Malvasio, V., & Danti, A. D. (2011). Effectiveness of primary endoscopic incision in treatment of ectopic ureterocele associated with duplex system. *Urology*, 77(1), 191–194 doi:10.1016/j.urology.2010.02.061.

Andrioli, V., Guerra, L., Keays, M., Keefe, D. T., Tang, K., Sullivan, K. J., …, & Leonard, M. P. (2018). Active surveillance for antenatally detected ureteroceles: predictors of success. *Journal of Pediatric Urology*, *14*(3), 243 doi:10.1016/j.jpurol.2018.02.012.

Beganovic, A., Klijn, A. J., Dik, P., & De Jong, T. P. (2007). Ectopic ureterocele: long-term results of open surgical therapy in 54 patients. *Journal of Urology*, *178*(1), 251–254 doi: 10.1016/j.juro.2007.03.063.

Castagnetti, M., Cimador, M., Sergio, M., & de Grazia, E. (2004). Transurethral incision of duplex system ureteroceles in neonates: does it increase the need for secondary surgery in intravesical and ectopic cases? *BJU International*, *93*(9), 1313–1317 doi:10.1111/j.1464-410X. 2004.04861.x.

Chertin, B., Rabinowitz, R., Pollack, A., Koulikov, D., Fridmans, A., Hadas-Halpern, I., …, & Farkas, A. (2005). Does prenatal diagnosis influence the morbidity associated with left in situ nonfunctioning or poorly functioning renal moiety after endoscopic puncture of ureterocele? *Journal of Urology*, *173*(4), 1349–1352 doi:10.1097/01.ju.0000155439.68182.b4.

Cohen, S. A., Juwono, T., Palazzi, K. L., Kaplan, G. W., & Chiang, G. (2015). Examining trends in the treatment of ureterocele yields no definitive solution. *Journal Pediatric Urology*, *11*(1), 29 e21–e26, doi:10.1016/j.jpurol.2014.09.001.

Cooper, C. S., Passerini-Glazel, G., Hutcheson, J. C., Iafrate, M., Camuffo, C., Milani, C., …, & Snyder, H. M., 3rd (2000). Long-term followup of endoscopic incision of ureteroceles: intravesical versus extravesical. *Jorunal of Urology*, *164*(3 Pt. 2), 1097–1099 discussion 1099-1100.

de Jong, T. P., Dik, P., Klijn, A. J., Uiterwaal, C. S., & van Gool, J. D. (2000). Ectopic ureterocele: results of open surgical therapy in 40 patients. *Journal of Urology*, *164*(6), 2040–2043 discussion 2043-2044.

Decter, R. M., Roth, D. R., & Gonzales, E. T. (1989). Individualized treatment of ureteroceles. *Journal of Urology*, *142*(2 Pt. 2), 535–537 discussion 542-533.

Decter, R. M., Sprunger, J. K., & Holland, R. J. (2001). Can a single individualized procedure predictably resolve all the problematic aspects of the pediatric ureterocele? *Journal of Urology*, *165*(6 Pt. 2), 2308–2310.

Dewan, P. A. (2000). Ureteric reimplantation: a history of the development of surgical techniques. *BJU International*, *85*(8), 1000–1006.

Gomes, J., Mendes, M., Castro, R., & Reis, A. (2002). Current role of simplified upper tract approach in the surgical treatment of ectopic ureteroceles: a single centre's experience. *European Urology*, *41*(3), 323–327.

Gran, C. D., Kropp, B. P., Cheng, E. Y., & Kropp, K. A. (2005). Primary lower urinary tract reconstruction for nonfunctioning renal moieties associated with obstructing ureteroceles. *Journal of Urology*, *173*(1), 198–201 doi:10.1097/01.ju.0000148374.64478.b5.

Gross, R. E., & Clatworthy, H. W., Jr. (1950). Ureterocele in infancy and childhood. *Pediatrics*, *5*(1), 68–77.

Hendren, W. H., & Mitchell, M. E. (1979). Surgical correction of ureteroceles. *Journal of Urology*, *121*(5), 590–597.

Holmes, N. M., Coplen, D. E., Strand, W., Husmann, D., & Baskin, L. S. (2002). Is bladder dysfunction and incontinence associated with ureteroceles congenital or acquired? *Journal of Urology*, *168*(2), 718–719.

Husmann, D. A., Ewalt, D. H., Glenski, W. J., & Bernier, P. A. (1995). Ureterocele associated with ureteral duplication and a nonfunctioning upper pole segment: management by partial nephroureterectomy alone. *Journal of Urology*, *154*(2 Pt 2), 723–726.

Jayanthi, V. R., Churchill, B. M., McLorie, G. A., & Khoury, A. E. (1995). Concomitant bladder neck closure and Mitrofanoff diversion for the management of intractable urinary incontinence. *Journal of Urology*, *154*(2 Pt 2), 886–888.

Jesus, L. E., Farhat, W. A., Amarante, A. C., Dini, R. B., Leslie, B., Bagli, D. J., …, & Pippi Salle, J. L. (2011). Clinical evolution of vesicoureteral reflux following endoscopic puncture in children with duplex system ureteroceles. *Journal of Urology*, *186*(4), 1455–1458 doi:10.1016/j.juro.2011.05.057.

Johnston, J. H., & Johnson, L. M. (1969). Experiences with ectopic ureteroceles. *British Jorunal of Urology*, *41*(1), 61–70.

Lee, Y. S., Im, Y. J., Shin, S. H., Bascuna, R. T., Jr., Ha, J. Y., & Han, S. W. (2015). Complications after common sheath reimplantation in pediatric patients with complicated duplex system. *Urology*, *85*(2), 457–462 doi:10.1016/j.urology.2014.10.014.

Lewis, J. M., Cheng, E. Y., Campbell, J. B., Kropp, B. P., Liu, D. B., Kropp, K., …, & Kaplan, W. E. (2008). Complete excision or marsupialization of ureteroceles: does choice of surgical approach affect outcome? *Jorunal of Urology*, *180*(4 Suppl.), 1819–1822 discussion 1822-1813. doi:10.1016/j.juro.2008.04.078.

Merguerian, P. A., Byun, E., & Chang, B. (2003). Lower urinary tract reconstruction for duplicated renal units with ureterocele. Is excision of the ureterocele with reconstruction of the bladder base necessary? *Jorunal of Urology*, *170*(4 Pt. 2), 1510–1513 discussion 1513. doi:10.1097/01.ju.0000084625.17209.e4.

Pohl, H. G. (2011). Recent advances in the management of ureteroceles in infants and children: why less may be more. *Current Opinion in Urology*, *21*(4), 322–327 doi:10.1097/MOU.0b013e328346d455.

Rich, M. A., Keating, M. A., Snyder, H. M., 3rd, & Duckett, J. W. (1990). Low transurethral incision of single system intravesical ureteroceles in children. *Jorunal of Urology*, *144*(1), 120–121.

Scherz, H. C., Kaplan, G. W., Packer, M. G., & Brock, W. A. (1989). Ectopic ureteroceles: surgical management with preservation of continence--review of 60 cases. *Jorunal of Urology*, *142*(2 Pt 2), 538–541 discussion 542-533.

Shekarriz, B., Upadhyay, J., Fleming, P., Gonzalez, R., & Barthold, J. S. (1999). Long-term outcome based on the initial surgical approach to ureterocele. *Jorunal of Urology*, *162*(3 Pt 2), 1072–1076.

Sherman, N. D., Stock, J. A., & Hanna, M. K. (2003). Bladder dysfunction after bilateral ectopic ureterocele repair. *Jorunal of Urology*, *170*(5), 1975–1977 doi:10.1097/01.ju.0000085654.46865.f0.

Shimada, K., Matsumoto, F., & Matsui, F. (2007). Surgical treatment for ureterocele with special reference to lower urinary tract reconstruction. *International Jorunal of Urology*, *14*(12), 1063–1067 doi:10.1111/j.1442-2042.2007.01903.x.

Shokeir, A. A., & Nijman, R. J. (2002). Ureterocele: an ongoing challenge in infancy and childhood. *BJU International*, *90*(8), 777–783.

Snyder, H. M., & Johnston, J. H. (1978). Orthotopic ureteroceles in children. *Jorunal of Urology*, *119*(4), 543–546.

Upadhyay, J., Bolduc, S., Braga, L., Farhat, W., Bagli, D. J., McLorie, G. A., …, & El-Ghoneimi, A. (2002). Impact of prenatal diagnosis on the morbidity associated with ureterocele management. *Jorunal of Urology*, *167*(6), 2560-2565.

Wang, M. H., Greenfield, S. P., Williot, P., & Rutkowski, J. (2008). Ectopic ureteroceles in duplex systems: long-term follow up and 'treatment-free' status. *Journal of Pediatric Urology*, *4*(3), 183–187 doi:10.1016/j.jpurol.2007.12.003.

Weiss, D. A. (2015). Management of ureterocele? The search for the holy grail. *Journal of Urology*, *193*(2), 398–399 doi:10.1016/j.juro.2014.11.077.

Yamazaki, Y., Setoguchi, K., Yago, R., Suzuki, M., & Toma, H. (1999). Common sheath reimplantation with ureteral plication: a useful technique for the management of ectopic ureterocele. *International Journal of Urology*, *6*(10), 532–535.

CHAPTER 11

Secondary surgery after failure of minimally invasive procedures

Kourosh Afshar, Andrew E. MacNeily
Department of Urologic Sciences, Division of Pediatric Urology BC Children's Hospital, University of British Columbia, Vancouver, BC, Canada

The management of ureteroceles has been a source of constant controversy. The variations in patient anatomy and their clinical presentation make the generalization of findings in different studies problematic. Authors have attempted to assess the effects of many variables on the outcomes of surgical treatment. To name a few, they include single versus duplicated ureter, intra versus extravesical location, level of renal function and clinical presentation. In addition, over the last few decades with a better understanding of the natural history of some urological abnormalities such as vesicoureteral reflux (VUR), minimally invasive and conservative managements have gained more popularity. The ideal management of ureteroceles would be minimally invasive, and prevent unfavorable outcomes with a minimum number of interventions.

Minimally invasive surgery for ureteroceles (MIS) can refer to transurethral puncture (TUP) or its variant watering can puncture (WCP), transurethral incision (TUI), or most recently laparoscopic clipping of the ureter subtending the ureterocele. Minimally invasive approaches have, at times been described as a temporizing maneuver, until patients are well enough for "definitive" reconstruction, as well as a first line treatment that works for most cases.

Indeed, endoscopic treatment first started as a temporizing measure to drain obstructed and infected collecting systems (Shokeir & Nijman, 2002). Later on, multiple studies emerged showing the value of endoscopic treatment as a definitive management (Adorisio et al., 2011; Chertin, Mohanan, Farkas, & Puri, 2007). Endoscopic treatment has the advantage of being minimally invasive with a short anesthetic time and no need for hospital admission in most cases. One method to gauge the success (or failure) of endoscopic treatment is the incidence of performing additional operations. However, there are divergent views in the literature as to who would be a

Ureteroceles
http://dx.doi.org/10.1016/B978-0-12-817888-1.00011-2

candidate for additional surgeries, hence the wide range of reported success rate from 0% to 100% (Byun & Merguerian, 2006). For example in many studies de novo or persistent vesicoureteral reflux has been considered an indication to intervene whether or not it is associated with infection (Gander, Asensio, Royo, & Lloret, 2016; Jesus et al., 2011). Studies that are more recent have shown asymptomatic VUR following endoscopic interventions for ureteroceles can be safely monitored without adverse clinical outcomes. This parallels the modern evolution in our treatment of primary VUR (Castagnetti, Cimador, Sergio, & de Grazia, 2004; Chertin et al., 2007). Postoperative VUR has been reported in 30%–100% of cases. In series that have attempted to follow these patients approximately 50%–70% of them resolved or downgraded spontaneously (Adorisio et al., 2011; Jesus et al., 2011).

Several studies have attempted to evaluate risk factors for endoscopic failure. For example in a systematic review of the studies up to 2005, Byun et al. found that extra vesical ureteroceles are 2.78 times more likely to have secondary surgeries (Byun & Merguerian, 2006). The corresponding risk ratios for duplex versus single systems and the presence of preoperative reflux were 3.93 and 1.56 respectively. The co-presence of these risk factors did not have an additive effect. Similar findings have been reported in more recent studies (Di Renzo, Ellsworth, Caldamone, & Chiesa, 2010; Sander et al., 2015).

Duplex ureteroceles have been reported to have a higher risk of febrile UTI post incision (Hodhod, Jednak, Khriguian, & El-Sherbiny, 2017). Postnatal presentation has also been associated with a higher likelihood of a secondary surgery (Hodhod et al., 2017).

The technical nuances of endoscopic treatments (e.g., puncture vs. incision) may also affect the outcome. Initial studies whereby large incisions were used resulted in a high incidence of de novo VUR. Different variations of the original endoscopic technique have been employed to reduce this unwanted outcome such as a distal incision of extra vesical ureteroceles and puncture as opposed to incision. Recently, Haddad et al. reported a failure rate of 38% for multiple punctures (watering can technique) versus 71% for incision, mainly due to the lower incidence of postoperative VUR (Haddad et al., 2017; Maizels, Liu, Gong, Kaplan, & Cheng, 2016).

Ligation or clipping of the ureter draining the upper pole is a more recently described MIS approach. This intervention is an extension of the (mostly) successful practice of ureteral ligation of poorly functioning native kidneys at the time of renal transplant in order to facilitate graft to host pyeloureterostomy. This operation is more suitable for poorly functioning moieties, which usually involute following ligation. There are no studies with large numbers of patients and long-term follow up in the non-transplant population. We also do not know the outcome of this surgery in the context of existing ureteroceles (Lopes et al., 2018).

Following failure of endoscopic treatment, several surgical interventions have been applied:

Repeat puncture

Repeat puncture has been employed in up to 36% of ureteroceles (Adorisio et al., 2011) and more commonly in ectopic (12%) versus intravesical ureteroceles (2%) with reported success rates of 100% (Jawdat, Rotem, Kocherov, Farkas, & Chertin, 2018). This treatment is mainly used when the intended decompression does not occur after the first puncture.

Ureteral reimplantation

Ureteral reimplantation, with or without resection or marsupialization of the ureterocele has the combined advantage of preserving renal function and the ability to manage preoperative or de novo VUR (Gander et al., 2016; Haddad et al., 2017; Paye-Jaouen et al., 2015). The disadvantages include technical challenges in young patients, and the post-operative pain and morbidity especially with an intravesical approach.

Endoscopic treatment of VUR

Endoscopic treatment of persistent or de novo VUR has also been reported as an effective measure in several studies (Adorisio et al., 2011; Haddad et al., 2017; Jesus et al., 2011), but the success rates are not clear.

Upper pole hemi-nephrectomy

Ablative procedures such as upper pole hemi-nephrectomy are mainly reserved for poorly functioning renal moieties. The major concern is the small risk of damage to the lower pole moiety of a duplex system, especially in younger children owing to damage or spasm of the small vessels to the normal parenchyma. Although this approach is highly successful in restoring normal urinary tract function, it does not deal with the issue of VUR and leaves a ureteral stump behind that can act as a nidus of infection (Michaud & Akhavan, 2017). Therefore, some urologists have used a combined upper and lower urinary tract approach following failed MIS management. There is no evidence that the outcome of hemi-nephrectomies following failed endoscopic treatments are different from when used as the primary approach.

Ipsilateral ureteroureterostomy (IUU)

IUU is well described for the management of ureteral duplication anomalies. There is a paucity of data regarding its use after failed endoscopic treatment of ureteroceles. Most reports have only a few cases with unclear follow up (Rodrigues, Estevao-Costa, & Fragoso, 2016).

Conclusions

No definitive surgical repair exists for ureteroceles because of the heterogeneous nature of this relatively rare condition. The main criticism of available studies pertaining to this chapter is the incomplete follow up and outcome reports after any secondary surgeries undertaken. Most studies do not go into the details of patient outcomes after a second or third surgery leaving the reader with the impression that these patients did well.

Secondary surgery rates following endoscopic puncture or incision seem to be primarily for de novo reflux either into the ureterocele moiety itself or into the ipsilateral lower pole. To confuse matters, many studies do not define de novo reflux as that which occurs into the ureterocele or into the ipsilateral lower pole. The occurrence of de novo reflux into the ureterocele appears to be lower with a watering can approach compared to a single puncture. However, the presence of VUR itself was historically an indication for secondary surgery. This has more recently been called into question, just as all indications for primary VUR have. The subjective and evolving nature of the indications for intervention in VUR makes it difficult to predict the absolute risk for secondary surgery following endoscopic surgery of ureteroceles.

That said it would appear that large ectopic ureteroceles, which are managed initially by endoscopic puncture or incision, are at greatest risk for needing secondary surgery (Table 11.1). Endoscopic clipping is a novel and innovative technique gleaned from the transplant experience. However, it requires further application and follow-up before any definitive recommendations can be made regarding this approach.

Table 11.1 The need for additional surgery related to classification of ureterocele.

	Number of cases	N (%) second surgery	N (%) Second surgery extravesical	N (%) second surgery intravesical	% Second surgery duplication	% Second surgery single ureter
Chowdhary, Kandpal, Sibal, Srivastava, and Vasudev (2017)	44	3/44 (7)	1/21 (5)	2/13 (15)		
Rodrigues (2016)	18	3/18(17)	0/8 (0)	3/10 (33)		
Boucher, Cloutier, Rousseau, Charrois, and Bolduc (2013)	66	18/66 (27)	3/30(10)	15/36 (42)		
Castagnetti (2004)	32	9/32 (28)	2/7 (28)	7/25 (28)		
Di Renzo (2010)	45	21/45 (47)	17/25 (63)	6/20 (33)	15/34 (44)	2/11 (18)
Castagnetti et al. (2013)	41	21/41 (51)	13/24 (54)	8/17 (47)		
Hagg et al. (2000)	60	19/60 (32)	14/29 (48)	5/22 (23)	19/51 (37)	0/9 (0)

References

Adorisio, O., Elia, A., Landi, L., Taverna, M., Malvasio, V., & Danti, A. D. (2011). Effectiveness of primary endoscopic incision in treatment of ectopic ureterocele associated with duplex system. *Urology*, 77(1), 191–194 doi:10.1016/j.urology.2010.02.061.

Boucher, A., Cloutier, J., Rousseau, A., Charrois, R., & Bolduc, S. (2013). Is an initial endoscopic treatment for all ureteroceles appropriate? *Journal of Pediatric Urology*, *9*(3), 339–343 doi:10.1016/j.jpurol.2012.04.009.

Byun, E., & Merguerian, P. A. (2006). A meta-analysis of surgical practice patterns in the endoscopic management of ureteroceles. *The Journal of Urology*, *176*(4 Pt 2), 1871–1877 discussion 1877. doi:S0022-5347(06)00601-X.

Castagnetti, M., Cimador, M., Sergio, M., & de Grazia, E. (2004). Transurethral incision of duplex system ureteroceles in neonates: does it increase the need for secondary surgery in intravesical and ectopic cases? *BJU International*, *93*(9), 1313–1317 doi:10.1111/j.1464-410X.;1; 2004.04861.x.

Castagnetti, M., Vidal, E., Burei, M., Zucchetta, P., Murer, L., & Rigamonti, W. (2013). Duplex system ureterocele in infants: should we reconsider the indications for secondary surgery after endoscopic puncture or partial nephrectomy? *Journal of Pediatric Urology*, *9*(1), 11–16 doi:10.1016/j.jpurol.2012.06.016.

Chertin, B., Mohanan, N., Farkas, A., & Puri, P. (2007). Endoscopic treatment of vesicoureteral reflux associated with ureterocele. *The Journal of Urology*, *178*(4 Pt 2), 1594–1597 doi:S0022-5347(07)00868-3.

Chowdhary, S. K., Kandpal, D. K., Sibal, A., Srivastava, R. N., & Vasudev, A. S. (2017). Ureterocele in newborns, infants and children: ten year prospective study with primary endoscopic deroofing and double J (DJ) stenting. *Journal of Pediatric Surgery*, *52*(4), 569–573 doi:S0022-3468(16)30297-4.

Di Renzo, D., Ellsworth, P. I., Caldamone, A. A., & Chiesa, P. L. (2010). Transurethral puncture for ureterocele-which factors dictate outcomes? *The Journal of Urology*, *184*(4 Suppl), 1620–1624 doi:10.1016/j.juro.2010.04.023.

Gander, R., Asensio, M., Royo, G. F., & Lloret, J. (2016). Evaluation of the initial treatment of ureteroceles. *Urology*, *89*, 113–117 doi:10.1016/j.urology.2015.11.025.

Haddad, J., Meenakshi-Sundaram, B., Rademaker, N., Greger, H., Aston, C., Palmer, B. W., …, & Frimberger, D. (2017). "Watering can" ureterocele puncture technique leads to decreased rates of de novo vesicoureteral reflux and subsequent surgery with durable results. *Urology*, *108*, 161–165 doi:S0090-4295(17)30621-0.

Hagg, M. J., Mourachov, P. V., Snyder, H. M., Canning, D. A., Kennedy, W. A., Zderic, S. A., & Duckett, J. W. (2000). The modern endoscopic approach to ureterocele. *The Journal of Urology*, *163*(3), 940–943 doi:S0022-5347(05)67857-3.

Hodhod, A., Jednak, R., Khriguian, J., & El-Sherbiny, M. (2017). Transurethral incision of ureterocele: does the time of presentation affect the need for further surgical interventions? *Journal of Pediatric Urology*, *13*(5), 501.e1-501.e6. doi:S1477-5131(17)30106-7.

Jawdat, J., Rotem, S., Kocherov, S., Farkas, A., & Chertin, B. (2018). Does endoscopic puncture of ureterocele provide not only an initial solution, but also a definitive treatment in all children? over the 26 years of experience. *Pediatric Surgery International*, *34*(5), 561–565 doi:10.1007/s00383-018-4258-9.

Jesus, L. E., Farhat, W. A., Amarante, A. C., Dini, R. B., Leslie, B., Bagli, D. J., …, & Pippi Salle F.J.L. (2011). Clinical evolution of vesicoureteral reflux following endoscopic puncture in children with duplex system ureteroceles. *The Journal of Urology*, *186*(4), 1455–1458 doi:10.1016/j.juro.2011.05.057.

Lopes, R. I., Fernandez, N., Koyle, M. A., Dos Santos, J., Perez, J., Romao, R., & Lorenzo, A. J. (2018). Clinical outcomes of the upper urinary tract after ureteral clipping for treat-

ment of low functioning or nonfunctioning renal moieties. *The Journal of Urology*, *199*(2), 558–564 doi:S0022-5347(17)77566-0.
Maizels, M., Liu, D., Gong, E. M., Kaplan, W. E., & Cheng, E.Y. (2016). Endoscopic ureterocele-otomy - retrograde incision from orifice (RIO) of urethral segment of ureterocele and extending to bladder neck is a feasible and simple procedure. *Journal of Pediatric Urology*, *12*(3), 137–138 doi:10.1016/j.jpurol.2016.05.010.
Michaud, J. E., & Akhavan, A. (2017). Upper pole heminephrectomy versus lower pole ureteroureterostomy for ectopic upper pole ureters. *Current Urology Reports*, *18*(3), 21-017-0664-0. doi:10.1007/s11934-017-0664-0.
Paye-Jaouen, A., Pistolesi, F., Botto, N., Enezian, G., Grapin-Dagorno, C., Peycelon, M., …, & El-Ghoneimi, A. (2015). Long-term bladder function after ureterocele decompression in children. *The Journal of Urology*, *193*(5 Suppl), 1754–1759 doi:10.1016/j.juro.2014.10.115.
Rodrigues, I., Estevao-Costa, J., & Fragoso, A. C. (2016). Complete ureteral duplication: Outcome of different surgical approaches. *Acta Medica Portuguesa*, *29*(4), 275–278 doi:10.20344/amp.6329.
Sander, J. C., Bilgutay, A. N., Irina, S., Koh, C. J., Nicolette, J., Gonzales, E. T., & Abhishek, S. (2015). Outcomes of endoscopic incision for the treatment of ureterocele in children at a single institution. *Journal of Urology*, *193*(2), 662–667 doi:10.1016/j.juro.2014.08.095.
Shokeir, A. A., & Nijman, R. J. (2002). Ureterocele: an ongoing challenge in infancy and childhood. *BJU International*, *90*(8), 777–783 doi:10.1046/j.1464-L 4096.2002.02998x.

CHAPTER 12

Long-term urological morbidity associated with ureterocele observational versus surgical approach

Luis Guerra[a], **Paul F. Austin**[b]
[a]Children's Hospital of Eastern Ontario, University of Ottawa, Ottawa, ON, Canada
[b]Baylor College of Medicine, Texas Children's Hospital, Houston, TX, United States

Ureterocele—Long-term urological morbidity associated with observational approach

Ureteroceles can be associated with many potential complications of the lower and upper urinary tracts. Several anatomical factors influence potential complications and morbidities associated with ureteroceles and these anatomical factors must be taken into consideration in the short and long-term care of the patient. The key anatomical features that influence these potential complications include the positioning of the ureterocele within the lower urinary tract and whether or not there is ureteral duplication present.

Anatomical factors—Single ureter versus ureteral duplication

Obstruction

Because of the dilation of the terminal ureter with ureterocele and associated poor drainage, it is no surprise that obstruction of the upper urinary tract may result from a ureterocele. This obstruction can occur in both a single system ureter and kidney or in a ureteral duplication with the associated upper pole ureter and kidney obstructed from the ureterocele. Treatment for obstructed ureterocele target decompression and the options include endoscopic puncture or surgical reconstruction. Not all ureteroceles, particularly single system ureteroceles, result in obstruction of the upper urinary tract. In the setting of low-grade or stable hydronephrosis with adequate drainage on diuretic renal scan; an observational approach without treatment can be taken in the select cases. These non-obstructing single system ureteroceles typically are small in

http://dx.doi.org/10.1016/B978-0-12-817888-1.00012-4

size. Another caveat where observation can be the best approach is in a poorly functioning kidney, such as a multicystic dysplastic kidney associated with ureterocele. Observation is recommended as most of these ureteroceles will involute along with the multicystic dysplastic kidney over their natural history without any complications or resultant morbidity (Coplen & Austin, 2004).

Stones

Urolithiasis or stones within the ureterocele occurs due to the stasis of the urine within the obstructed ureterocele. This allows supersaturation of the urine with precipitation of the urinary crystals and subsequent stone formation. The association of urolithiasis may occur in single systems (Gilbert, Hernanz-Schulman, & Pope, 2001; Gupta, Gupta, Khan, & Basu, 2017) or in duplicated ureteral systems (Gharbi, Gargah, Boukesra, Hellal, & Sayed, 2013). Stones within ureteroceles have been reported in orthotopic located ureteroceles as well as ectopic ureteroceles (Gharbi et al., 2013; Gilbert et al., 2001; Gupta et al., 2017). The presence of urolithiasis within a ureterocele has been reported not only in adults (Shah et al., 2008) but also during infancy (Gilbert et al., 2001). Treatment for stones within ureteroceles is generally performed endoscopically with holmium laser (El-Khoury, Aw, Gilbourd, & Liu, 2018; Shah et al., 2008) but others have also recommended open removal in complex anatomy requiring lower urinary tract reconstruction (Dominici, Travaglini, Maleci, di Cello, & Rizzo, 2003; Gharbi et al., 2013).

Stones that are associated with ureteroceles may present in confusing manners. Large stones within ureteroceles are often initially diagnosed as bladder stones (Dominici et al., 2003; El-Khoury et al., 2018). Ureterocele stones have even been associated with urinary retention when they are located within a prolapsed ureterocele (Gupta et al., 2017; Scovell, Chan, & Khavari, 2014).

Urinary tract infections

Similar to obstruction and stones the pathogenesis for increased risk of urinary tract infections associated with ureteroceles is due to the urinary stasis from poor drainage out of the ureterocele. Decompression of the obstructed ureterocele results in diminished risk of urinary tract infection and can be accomplished via endoscopic or open surgical treatment. In the setting of stable renal function or even poor renal function, treatment with prophylactic antibiotics has been advocated for the reduction of urinary tract infections when future treatment is under consideration (Husmann, Ewalt, Glenski, & Bernier, 1995).

Vesicoureteral reflux

Vesicoureteral reflux (VUR) is generally associated with ureteroceles in the setting of duplicated ureters. VUR is present in the lower pole ureter rather than the upper pole ureter that has the ureterocele. The reflux is attributed to the superior and lateral insertion site of the lower pole ureter in the bladder with less submucosal and intramural tunneling but may also be associated with distortion of the bladder by the ureterocele (Coplen & Duckett, 1995; Husmann et al., 1995).

Renal function

Renal function can be affected in the setting of ureteroceles particularly when they are associated with obstruction. As mentioned earlier, the urinary stasis associated with ureterocele will promote urolithiasis as well as urinary tract infections that can further add insult to already compromised renal function. Impaired renal function is more commonly seen in upper renal pole segments of ureteral duplications with ectopic ureteroceles, as these are more often obstructed.

Table 12.1 Long-term urological morbidity associated with observational approach.

Key points
There are few studies specifically examining long-term bladder function associated with the natural history of ureteroceles and/or ureterocele treatments
Bladder dysfunction is relatively common with ectopic ureteroceles associated with ureteral duplications
Treatments are mostly successful with the majority needing only behavioral modification and standard urotherapy
Surgical intervention does not seem to play a significant role in the long-term bladder function outcome
Upper versus lower urinary tract intervention similarly does not seem to be a key determinant of bladder function.

Anatomical factors—Ureterocele location

Obstruction

The anatomical location of the ureterocele is a key determinant towards defining the involved, obstructed urinary tract segment. For instance, an orthotopic ureterocele will only involve the ureter and kidney whereas an ectopic ureterocele in a ureteral duplication may also involve the lower urinary tract. Lower urinary tract obstruction is a result of the ureterocele obstructing the bladder neck or prolapsing across the bladder neck and down the urethra (Coplen & Duckett, 1995; Husmann et al., 1995; Shimada, Matsumoto, & Matsui, 2007).

Bladder dysfunction

Bladder dysfunction is anatomically associated with ectopic ureteroceles that are associated with ureteral duplication. Urinary incontinence may result from the distortion of the bladder neck due to widening as a direct result from the ureterocele. Urinary incontinence may be indirectly caused by the ureterocele due to bladder outlet obstruction that results in architectural changes in the bladder wall that lead to overactivity and poor storage or myogenic failure with poor emptying. Another potential concern is the impact of surgical treatment for ureteroceles that may result in urinary incontinence after ureterocele resection (Coplen & Duckett, 1995; Husmann et al., 1995; Mor, Ramon, Raviv, Jonas, & Goldwasser, 1992).

In the first report examining bladder function with urodynamics, Abrahamsson et al examined 32 patients with ectopic ureteroceles by performing noninvasive and invasive urodynamics to characterize their bladder function (Abrahamsson, Hansson, Sillen, Hermansson, & Hjalmas, 1998). The patients underwent a variety of upper or lower urinary tract focused procedures. Overall, the most common bladder dysfunction symptom was infrequent voiding (59%) and urinary incontinence was only present in 9%. Urinary tract infections were reported in 5 patients (16%) and 2 patients (6%) required clean intermittent catheterization. The patients exhibited normal detrusor compliance but the average bladder capacity was 160% of normal for age. 60% of the patients that underwent bladder surgery versus 33% that did not have bladder surgery showed an increase in bladder capacity. Interestingly, the highest bladder capacity was seen in the youngest children but those children older than 7 years had a normal bladder volume. Majority of patients had a normal flow curve with a mean maximum flow rate of 26 mL/s (median 25, range 10.3–49.6). Significant post void residuals were defined as a volume exceeding 5 mL and were seen in approximately half of the patients and did not differ in the cohort that had bladder surgery versus no bladder surgery. Finally, neither increased bladder capacity nor residual urine was significantly correlated with ureterocele size, extension into the urethra, VUR before surgery, or ureterocele resection. The authors concluded that the bladder dysfunction associated with ectopic ureterocele does not seem to be the result of surgery but an integral part of the disorder (Abrahamsson et al., 1998).

In a multi institutional, retrospective review, Holmes et al reviewed all types of ureteroceles to characterize the patterns of bladder dysfunction and its association with prior surgical treatments (Holmes, Coplen,

Strand, Husmann, & Baskin, 2002). Out of 616 patients, 39 patients were identified that had bladder dysfunction and all were associated with ectopic ureteroceles in the setting of ureteral duplication. Urinary urgency and urinary incontinence were the most common bladder dysfunction symptoms and seen in 85% of the cohort. Infrequent voiding was seen in a minority of patients (13%). Urodynamics was performed in 34/39 patients and detrusor overactivity was the most common finding seen and correlated with their bladder dysfunction symptoms. Interestingly, stress urinary incontinence was seen in 6/34 patients (18%). Reduction of detrusor compliance was seen in 23% of the patients. The treatment of the latter dysfunction was identical to patients without ectopic ureteroceles and consisted of behavioral modification and pharmacologic management as indicated. Approximately 90% of the patients responded to the traditional nonsurgical management of urinary incontinence. There were four patients that underwent surgical intervention for their incontinence and two patients had placement of an artificial urinary sphincter and two patients underwent bladder neck closures. This multi institutional study did not identify any particular surgical approach that resulted in bladder dysfunction as patients treated with upper tract surgery alone had similar rates of incontinence to lower urinary tract surgery alone. Subsequently, they concluded that bladder dysfunction in patients with ectopic ureteroceles is congenital as opposed to acquired from surgical intervention (Holmes et al., 2002).

Table 12.2 Long-term complications with surgical approach.

Key points

- Extravesical ureterocele is not a good indication for incision due to its complications
- The "Windsock phenomenon" is the most feared complication post an endoscopic ureterocele incision
- With the marsupialization, the ureterocele unroofing incision must be extended down into the urethra in order to avoid windsock phenomenon
- Asymptomatic de-novo VUR post ureterocele incision may be followed conservatively
- The surgery for ureteric stump removal can be challenging due to the severe peri-ureteric fibrosis and adhesion to surrounding organs and major vessels usually caused by chronic infected stumps
- The vas deferens is closely related to the distal ureter and may be affected or injured by dissection during surgery for ureteric reimplantation for ectopic ureter or ureterocele in males

Further studies examining bladder function differ on assessment of the impact of total reconstructive bladder surgery on the urodynamic behavior of the bladder. Sherman et al identified 12 patients with bilateral ectopic ureteroceles that underwent lower urinary tract reconstruction after a variety of initial treatment such as endoscopic treatment in 9 patients and complete reconstruction in 3 patients (Sherman, Stock, & Hanna, 2003). They noted bladder dysfunction with incomplete bladder emptying in 70% of their patients as defined by post void residual greater than 20% of expected bladder capacity. Urodynamics was only performed in three patients and they noted elevated detrusor compliance with large postvoid residuals. Two patients required clean intermittent catheterization for continence (Sherman et al., 2003).

The group from Utrecht examined the long-term bladder function outcomes of 54 children that were treated for ectopic ureteroceles with total upper and lower urinary tract reconstruction (Beganovic, Klijn, Dik, & De Jong, 2007). The majority the patients were younger than 1 year old and the age of surgery ranged from 0-8.8 years (median 1.0). Follow-up was 2.3–15.6 years (median 9.6). Overall continence was noted in 94% of the patients. Among these patients, 11% were noted to have urinary frequency and 2% were noted to have infrequent voiding. In the 3 patients (6%) that were incontinent, all were girls that had a short and insufficient urethra based on a large cecoureterocele. Dysfunctional voiding documented by uroflow EMG was seen in 11 patients (20%) and these patients were treated with biofeedback. Additionally, approximately 20% of the patients experienced 1 or 2 uncomplicated episodes of urinary tract infection. The authors concluded that the vast majority of patients treated with total reconstructive bladder surgery become continent and do not suffer from lower urinary tract symptoms during the long-term (Beganovic et al., 2007).

A subsequent report that complements the above Utrecht study is the long-term outcomes after ureterocele decompression as the main or initial surgical treatment rather than complete urinary tract reconstruction (Paye-Jaouen et al., 2015). Paye-Jaouen et al. identified 45 patients that completed LUT (lower urinary tract) symptom questionnaires as well as had noninvasive urodynamics. In their study, they noted that most common LUT symptoms were infrequent urination, urinary urgency, incontinence and dysuria. The majority of the patients responded well to behavioral modification and pharmacologic therapy as indicated. An interesting finding that was seen in the ureterocele decompression group was the fact that 22% of the patients had significant uroflowmetry anomalies such as a significant

PVR and an activated EMG during voiding but none had overt symptoms of bladder dysfunction. The overall incidence of major bladder dysfunction was relatively low at only 3 cases (6%) of which only one needed CIC (2%) (Paye-Jaouen et al., 2015).

Ureterocele—long-term complications with surgical approach

Both the surgical or conservative ureterocele management aim to avoid recurrent UTIs, renal scarring, and to achieve renal function preservation. The surgical management of ureterocele can be challenging due to the spectrum of the uretero-vesical anatomy. Ureterocele repair can have possible different postoperative complications due to the presence of either a single system ureterocele or duplex system ureterocele and its anatomic differences. To minimize this, the precise upper and lower urinary anatomy should be assessed before any surgical treatment and this may be done through imaging tests such as VCUG, ultrasound or possibly a MRI. Renal scan may be used to assess function of the upper pole in duplex system ureterocele. CT scan should be avoided as much as possible due to the obvious radiation involved. Cystoscopy can determine the type, size, and location of the ureterocele on the trigone and rule out ectopic ureteric insertion.

Complications post ureterocele incision

When a single or duplex system intravesical ureterocele is obstructive, an endoscopic incision is recommended as initial management in order to alleviate the obstruction. Depending on the size of the incision or the bladder capacity, it may cause different degrees of VUR. In contrast to the observation that most patients with single system may require less secondary surgery (Di Renzo, Ellsworth, Caldamone, & Chiesa, 2010), duplex system ureterocele will usually need a second operation due to secondary reflux (Sander et al., 2015). Extravesical ureterocele is not a good indication for incision due to its complications, but can be initially incised in the presence of bladder neck obstruction and sepsis, and definitive reconstructive treatment has been shown to offer better results in these cases (Shekarriz, Upadhyay, Fleming, Gonzalez, & Barthold, 1999).

The "Windsock phenomenon" is the most feared complication post an endoscopic ureterocele incision and this is seen in cases of ectopic insertion of the ureterocele into the urethra, or cecoureterocele, in which the distal part of the ureterocele runs parallel to the urethra. During the endoscopy for incision, the location of the orifice of the ureterocele should be carefully determined to avoid this complication. If the ureterocele orifice is located ectopically inside the urethra, an endoscopic incision of the portion of the dilated ureterocele inside the bladder can create distal severe obstruction that resembles posterior urethral valve in boys. The anatomic "windsock" configuration of the intra-urethral segment of the ureterocele is precipitated by the urine flowing into the proximal ureterocele incision. This is a widely recognized complication that can lead to severe obstruction and have repercussion to the upper tract. If windsock phenomenon is suspected after endoscopic incision, a VCUG should be done to confirm it and an open surgical reconstruction with unroofing of the segment inside the urethra is indicated.

Despite the goal of increasing the drainage of the ureterocele system, the endoscopic incision should not be too large in size in order to avoid "de novo" occurrence of high-grade reflux to the upper moiety, which can lead to recurrent UTIs and need for reoperation. In general, VUR post transurethral ureterocele incision in duplex system has a higher reoperation rate compared to single system ureterocele. In a group of 51 children with 53 ureteroceles, the presence of de-novo VUR into the ureterocele after incision was 44% in single system ureterocele and 23% in duplex system ureterocele ($P = 0.22$) (Hodhod, Jednak, Khriguian, & El-Sherbiny, 2017). In this series, de-novo reflux into ureterocele after endoscopic incision carried a higher risk of surgical intervention, which was necessary in 51% of duplex system ureterocele and 35.7% of single system ureterocele. In contrast with these findings, Jesus et al. reported 60 patients with a mean age of 12 months who underwent transurethral puncture and found a new onset of vesicoureteral reflux postoperatively in 24 patients (40%), mainly ipsilateral to the lower pole or contralateral, and observed spontaneous resolution or improvement in 16 of 22 patients (72%); they recommend close surveillance as initial management (Jesus et al., 2011).

With incision for decompression, the presence of VUR preoperatively may impact the subsequent need for surgery. In a report of 41 neonates undergoing transurethral incision of a duplex system ureterocele, 51 % (21/41) required secondary surgery, but this was more common on VUR present before the incision rather than the occurrence of de novo VUR on the incised unit (Castagnetti, Cimador, Sergio, & De Grazia, 2004).

Technical variations of this incision have been reported in an attempt to avoid postoperative reflux, such as, "Watering can technique" (Palmer, Greger, Mannas, Kropp, & Frimberger, 2011) and low transverse incision, puncture with Laser and Bugbee electrode (Chertin, Fridmans, Hadas-Halpren, & Farkas, 2001).

In duplex system ureterocele, the length of the submucosal channel of the lower moiety ureter is known to be shorter and prone to VUR. However the orifice of the lower pole ureter may be slightly compressed by a large ureterocele, avoiding reflux to the lower pole or obstructing it (Fernbach, Feinstein, Spencer, & Lindstrom, 1997). The incision of the distal ureterocele may disturb the trigonal anatomy, alleviating the "pseudo obstruction" of the lower pole ureteric orifice and cause de novo VUR to the lower pole postoperatively, which in some cases can be quite severe. In a series of 6 and 15 patients with intravesical and extravesical duplex system ureteroceles and no VUR to the lower pole, the difference in rates of de novo lower pole VUR between these two groups was not statistically significant (Sander et al., 2015). Asymptomatic de novo VUR post ureterocele incision may be followed conservatively.

Time for ureterocele decompression in duplex system extravesical ureterocele

A duplex extravesical ureterocele can be associated with increased morbidity and the decision to perform an early endoscopic decompression is controversial. Husmann et al. reported on 72 infants younger than 2 weeks with an extravesical ureterocele associated with a duplex upper pole moiety who received either immediate endoscopic puncture of the ureterocele followed by antibiotic prophylaxis (32 patients) or antibiotic prophylaxis with plans for delayed surgical intervention (40 patients). He found no difference on the rate of UTI and progressive hydronephrosis between these 2 groups (Husmann, Strand, Ewalt, & Kramer, 2002). It seems that the decision to decompress the ureterocele should be dictated individually depending on the clinical scenario.

Complications post upper and lower urinary tract reconstruction

Complete reconstruction of the urinary tract can be used in the management of ureteroceles. This is done with an upper pole hemi-nephrectomy, double ureteric reimplantation, and in some cases excision of the posterior ureterocele mucosa with detrusor approximation may be necessary. However, despite good results most children will have resolution of the obstruction or reflux with only one of two of these procedures (Peters and Mendelsohn, in press). Alternatively, upper pole hemi-nephrectomy is indicated for non- or poorly-functioning upper pole moiety, and lower urinary tract reconstruction is done for large ureteroceles with good function of the upper pole.

Complications post upper pole hemi-nephrectomy

Ureteric stump syndrome and ischemia to the ureter

During an upper pole hemi-nephrectomy, the ligation and transection of the distal ureter should be done at the lowest possible level to avoid long distal dilated ureteric stump. However this may be challenging to accomplish through either a lumbar or laparoscopic approach. One must avoid extensive dissection separating both ureters as this may cause ureteric ischemia and secondary stricture. The common limitation in achieving a complete ureteric excision will result in a residual distal ureteric stump and ureterocele that can be a cause for recurrent infection, pus accumulation or stone formation. In general stones are more common in older patients and adults with single system ureterocele (Fig. 12.1), however, they can occur in duplex system ureteroceles and ureteric stumps.

VUR to the stump increases the chance of infection and other complications. The surgery for ureteric stump removal can be challenging due to

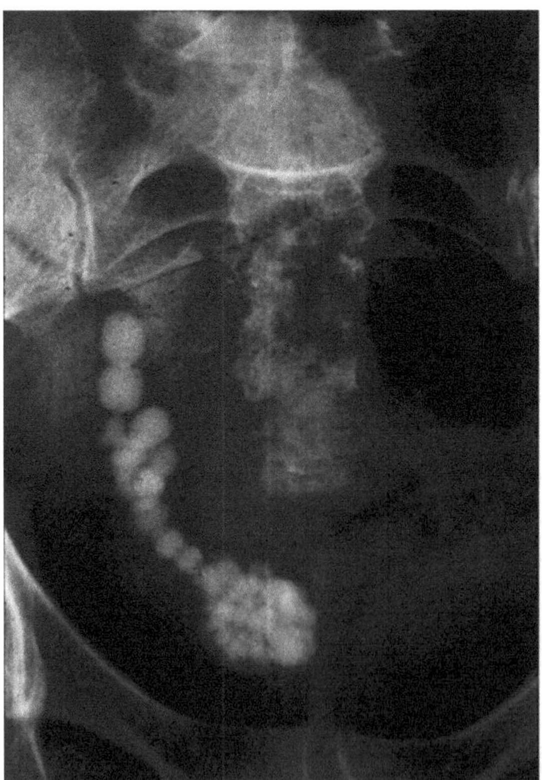

Figure 12.1 ***Multiple stones in an older patient with single system ureterocele.*** *Courtesy Dr. Rocha Brito.*

severe peri-ureteric fibrosis and adhesion to surrounding organs and major vessels caused by chronic infected stumps. In such cases the stump may be adherent to surrounding tissues and important structures including bowel and inferior vena cava. If during the surgery, severe fibrosis is present precluding the removal of ureteric stump, then marsupialization may be an option.

Vascular injury to the lower pole and subsequent decrease function

The two units of a duplex kidney share their blood supply by means of bifurcation of the artery and vein coming from the aorta and inferior vena cava respectively. Care must be taken to identify the upper pole vessels before ligation. A vascular pedicle bifurcation to the lower pole may be inadvertently ligated and transected during the upper pole heminephrectomy, which can cause ischemia and decreased renal function. The most dramatic situation is a complete loss of the lower pole renal function. For this reason, a careful dissection must be carried at the time of ligation of the upper pole vessels, and if a question resides about the direction of a specific vessel, a "bulldog clamp" can be used to assess for any patchy areas of ischemia on the lower pole. Ischemia to the lower pole can be also result from harsh or prolonged retraction of the vessels during the dissection.

Complications post reconstruction of the lower urinary tract

The surgical management post ureterocele incision failure is controversial, however in a survey of 233 pediatric urologists, most respondents advocated for ureteric reimplantation to manage reflux to the upper pole after endoscopic puncture, while some preferred endoscopic Deflux injection (Merguerian, Taenzer, Knoerlein, McQuiston, & Herz, 2010).

Depending on the size, the ureterocele associated with the upper pole ureter may be either marsupialized to the bladder cavity or completely excised including its posterior wall mucosa. With the marsupialization, the ureterocele unroofing incision must be extended down into the urethra in order to avoid windsock phenomenon. One must remember that the posterior bladder wall behind the ureterocele is usually thin, weaker and lacks adequate detrusor muscle support. When the posterior ureterocele mucosa is excised, a careful reconstruction of the posterior wall of the bladder trigone should be performed, sometimes involving a limited bladder neck plasty. This step may prevent postoperative detrusor weakness, bladder dysfunction and diverticulum formation.

As previously noted, the ureters should not be separated from the common sheath in order to avoid ischemic complications and late ureteric stricture.

Extra care should be taken to avoid injury of the contralateral ureter during the posterior detrusor muscle re-approximation, and the placement of a ureteric catheter in the contralateral ureter may prevent this complication. When the contralateral ureteral orifice is not identified in the bladder during an open approach, an ectopic ureter to the lower bladder neck or urethra should be suspected, and this can be confirmed with a cystoscopy.

When suturing the new submucosal tunnel to the ureteral reimplant, one must be careful to not pierce the ureteric wall along the tunnel with the suturing needle, which can result on a direct fistula between the reimplanted ureter segment and the bladder, with consequent iatrogenic VUR. It is important to firmly anchor the ureter to the detrusor muscle, which will avoid disruption of the implant with a "free ureter" delivering urine to the retro-vesical or intraperitoneal space, leading to large urinoma or ascites.

When properly done, either ureterocele excision or marsupialization may have acceptable outcomes. In a retrospective review of 3 academic institutions, 33 complete excision and 24 marsupialization of the ureteroceles in children with similar mean age and gender, found no difference in outcomes at 55 and 38 months of follow up in terms of vesicoureteral reflux (13% and 14%), bladder diverticulum (3% and 9%), continence (100% and 95%), urinary tract infection (43% and 46%) or voiding dysfunction (24% and 25%), respectively. (Lewis et al., 2008).

The most important step of the lower urinary tract reconstruction is the ureteric re-implantation. Occasionally, a ureterocele may be associated with a large dilated ureter that may need to be tapered to accommodate the ureter under the mucosal tunnel during the creation of a "flap-valve" at the UVJ. It is known that as a result of tapering, the ureteric wall may become thick and inelastic, which can lead to malfunction of the created UVJ valve, causing temporary obstruction or VUR. Because the reimplanted tapered ureteric wall will soften up over the months, the flap valve may take 6 months for ideal functionality.

The vas deferens is closely related to the distal ureter and may be affected or injured by dissection during surgery for ureteric reimplantation for ectopic ureter or ureterocele in males. The ureter is crossed anteriorly by the medial umbilical ligament and passes under the vas deferens before entering the bladder (Reed & Dodson, 1958). The ureteric dissection should be carried carefully in order to avoid transection of the vas or iatrogenic injury. As previously stated, the ureteric common sheath shares the same blood supply and dissecting them apart during the reimplantation may compromise the ureteral perfusion and also increase chance of vas injury and thus should be avoided.

Bladder dysfunction and surgery

The lack of adequate support at the posterior wall of the ureterocele can predispose to bladder dysfunction due to an asymmetric detrusor wall contraction during voiding and inadequate bladder neck function. This is in part caused by the outpouching of the bladder mucosa through the ureterocele floor at the level of the detrusor, which decreases the bladder contraction efficiency depending on the size of the floor weakness. This bladder dysfunction may be transient and improve over time, or resolve after bladder neck plasty. However in some cases of bilateral large ureteroceles with poor outcome after lower urinary tract reconstruction with or without posterior detrusor wall reinforcement, long-term bladder catheterization may be needed for incomplete emptying and large post void volumes. Some patients with associated small bladders may progress to bladder augmentation.

Paye-Jaouen et al. reviewed a group of 45 toilet-trained children treated for duplex system ureterocele with a mean follow up of 9.5 years (range 3–20), initial surgery was done at mean age of 5.1 months (range 6 days to 48 months). They were divided in Group 1 – (Ureterocele decompression with endoscopic incision, upper pole partial nephrectomy and ureteropelvic anastomosis), and Group 2 – (Primary or secondary bladder surgery: ureterocelectomy, ureterovesical reimplantation and bladder floor reconstruction). Voiding dysfunction was assessed by the Dysfunctional Voiding Scoring System questionnaire and noninvasive studies (Uroflowmety, EMG, and PVR). Urodynamics were done for abnormal Dysfunctional Voiding Scoring System or abnormal noninvasive studies. They concluded that ureterocele decompression alone in early childhood does not lead to major bladder dysfunction at long-term evaluation and showed that bladder dysfunction is rare in these patients even if secondary bladder surgery is needed (Paye-Jaouen et al., 2015).

Complications post ureteroureterostomy and ureteropyelostomy

The diversion of the urine from the upper moiety to the lower moiety pelvis or ureter can be achieved by either a ureteropyelostomy or a ureteroureterostomy. With this approach, the upper pole drainage is diverted to the pelvis or lower pole ureter and consequently the distal ureterocele should collapse. As the ureter from the upper pole is usually dilated, the anastomosis to the lower pole pelvis is easier to perform and carries lower complication rate. However, the anastomosis between both ureters can be challenging due to the disproportion in size between them, with the upper pole ureter usually being much larger, which may lead to complications such as urinary leak and stricture of the anastomosis. Whenever possible, a ureteric stent left for few weeks may prevent such complications. Leak of urine and urinomas around the ureteric anastomosis should be drained early to avoid the resultant fibrosis and constriction around the area. When deciding for this option of management, one must think that you may be transferring a potential problem from a dilated upper unit to a healthy and normal lower moiety. Despite this theoretical concern, ipsilateral ureteroureterostomy has been shown to be safe and efficacious for treating abnormalities associated with ureteral duplication and a significant discrepancy in ureteral size does not seem to preclude ureteroureterostomy (Lashley, McAleer, & Kaplan, 2001).

Complications post laparoscopic upper moiety ureteric clipping

A more recent management of the duplex kidney with ectopic ureter and poorly functioning upper pole moiety has been described, and it is done by clipping the upper pole ureter by an open or laparoscopic approach. Romao et al. described nine girls with ectopic ureters and non-functioning upper pole unit who underwent laparoscopic ligation and eight developed some degree of asymptomatic upper pole hydronephrosis (Romao et al., 2014). However, in some cases the upper pole may continue to dilate or be a potential focus for infection requiring a second intervention.

Subsequently, Lopes et al. reported that such technique can also be used in large ureterocele associated with non-functioning upper pole units (Lopes, Mello, Koyle, & Lorenzo, 2019). It seems that patient selection, including upper moieties that have minimum or no urine output, is the main factor in avoiding complication.

References

Abrahamsson, K., Hansson, E., Sillen, U., Hermansson, G., & Hjalmas, K. (1998). Bladder dysfunction: an integral part of the ectopic ureterocele complex. *The Journal of Urology*, *160*(4), 1468–1470.

Beganovic, A., Klijn, A. J., Dik, P., & De Jong, T. P. (2007). Ectopic ureterocele: long-term results of open surgical therapy in 54 patients. *The Journal of Urology*, *178*(1), 251–254 doi:10.1016/j.juro.2007.03.063.

Castagnetti, M., Cimador, M., Sergio, M., & De Grazia, E. (2004). Transurethral incision of duplex system ureteroceles in neonates: Does it increase the need for secondary surgery in intravesical and ectopic cases? *BJU International, 93*(9), 1313–1317 https://doi.org/10.1111/j.1464-410X.2004.04861.x.

Chertin, B., Fridmans, A., Hadas-Halpren, I., & Farkas, A. (2001). Endoscopic puncture of ureterocele as a minimally invasive and effective long-term procedure in children. *European Urology, 39*(3), 332–336 https://doi.org/10.1159/000052463.

Coplen, D. E., & Austin, P. F. (2004). Outcome analysis of prenatally detected ureteroceles associated with multicystic dysplasia. *The Journal of Urology, 172*(4 Pt 2), 1637–1639 discussion 1639. doi:10.1097/01.ju.0000138380.77747.7d.

Coplen, D. E., & Duckett, J. W. (1995). The modern approach to ureteroceles. *The Journal of Urology, 153*(1), 166–171 doi:10.1097/00005392-199501000-00068.

Di Renzo, D., Ellsworth, P. I., Caldamone, A. A., & Chiesa, P. L. (2010). Transurethral puncture for ureterocele-which factors dictate outcomes? *The Journal of Urology, 184*(4 Suppl), 1620–1624 https://doi.org/10.1016/j.juro.2010.04.023.

Dominici, A., Travaglini, F., Maleci, M., di Cello, V., & Rizzo, M. (2003). Giant stone in a complete duplex ureter with ureterocele. A case report. *Urologia Internationalis, 71*(3), 336–337 doi:10.1159/000072691.

El-Khoury, H. J., Aw, H. C., Gilbourd, D., & Liu, M. I. (2018). A large stone within a ureteroceles: a diagnostic pitfall and the utility of holmium laser deroofing as a viable surgical option. *Journal of Endourology Case Report, 4*(1), 18–20 doi:10.1089/cren.2017.0139.

Fernbach, S. K., Feinstein, K. A., Spencer, K., & Lindstrom, C. A. (1997). Ureteral duplication and its complications. *Radiographics, 17*(1), 109–127.

Gilbert, W. B., Hernanz-Schulman, M., & Pope, J. C. T. (2001). Development of small calculi in an infant with bilateral single system ureteroceles. *The Journal of Urology, 166*(5), 1860–1861.

Gharbi, Y., Gargah, T., Boukesra, T., Hellal, Y., & Sayed, S. (2013). Ureterocele containing a stone in a duplex system. *Saudi Journal of Kidney Diseases and Transplantation, 24*(5), 984–987.

Gupta, R., Gupta, S., Khan, D., & Basu, S. (2017). Bilateral single system orthotopic ureterocele with bilateral multiple calculi presented with retention of urine—an urological emergency. *Journal of Clinical and Diagnosis Research, 11*(7), PD01-PD02. doi:10.7860/JCDR/2017/28054.10134.

Hodhod, A., Jednak, R., Khriguian, J., & El-Sherbiny, M. (2017). Transurethral incision of ureterocele: does the time of presentation affect the need for further surgical interventions? *Journal of Pediatric Urology, 501*, e1-501.e6 https://doi.org/10.1016/j.jpurol.2017.02.016.

Holmes, N. M., Coplen, D. E., Strand, W., Husmann, D., & Baskin, L. S. (2002). Is bladder dysfunction and incontinence associated with ureteroceles congenital or acquired? *The Journal of Urology, 168*(2), 718–719.

Husmann, D. A., Strand, W. R., Ewalt, D. H., & Kramer, S. A. (2002). Is endoscopic decompression of the neonatal extravesical upper pole ureterocele necessary for prevention of urinary tract infections or bladder neck obstruction? *The Journal of Urology, 167*(3), 1440–1442 https://doi.org/10.1016/S0022-5347(05)65339-6.

Husmann, D. A., Ewalt, D. H., Glenski, W. J., & Bernier, P. A. (1995). Ureterocele associated with ureteral duplication and a nonfunctioning upper pole segment: management by partial nephroureterectomy alone. *The Journal of Urology, 154*(2 Pt 2), 723–726.

Jesus, L. E., Farhat, W. A., Amarante, A. C. M., Dini, R. B., Leslie, B., Bgli, D. J., …, & Pippi Salle, J. L. (2011). Clinical evolution of vesicoureteral reflux following endoscopic puncture in children with duplex system ureteroceles. *The Journal of Urology* https://doi.org/10.1016/j.juro.2011.05.057.

Lashley, D. B., McAleer, I. M., & Kaplan, G. W. (2001). Ipsilateral ureteroureterostomy for the treatment of vesicoureteral reflux or obstruction associated with complete ureteral duplication. *The Journal of Urology, 165*(2), 552–554 2001.

Lewis, J. M., Cheng, E.Y., Campbell, J. B., Kropp, B. P., Liu, D. B., Kropp, K., ..., & Kaplan, W. E. (2008). Complete excision or marsupialization of ureteroceles: does choice of surgical approach affect outcome? *Journal of Urology, 180*(4 Suppl), 1819–1822 https://doi.org/101016/j.juro.2008.04.

Lopes, R. I., Mello, M. F., Koyle, M. A., & Lorenzo, A. J. (2019). Ureteral clipping for the treatment of a non-functioning upper kidney moiety associated with a massive ureterocele step-by-step description of a novel technique. *Journal of Pediatric Urology, 15*(3), 284–285 https://doi.org/10.1016/j.jpurol.2019.03.005.

Merguerian, P. A., Taenzer, A., Knoerlein, K., McQuiston, L., & Herz, D. (2010). Variation in management of duplex system intravesical ureteroceles: a survey of pediatric urologists. *The Journal of Urology, 184*(4s), 1625–1630 https://doi.org/10.1016/j.juro.2010.03.113.

Mor, Y., Ramon, J., Raviv, G., Jonas, P., & Goldwasser, B. (1992). A 20-year experience with treatment of ectopic ureteroceles. *The Journal of Urology, 147*(6), 1592–1594 doi:10.1016/s0022-5347(17)37637-1.

Palmer, B. W., Greger, H., Mannas, D. B., Kropp, B. P., & Frimberger, D. (2011). Comparison of endoscopic ureterocele decompression techniques. Preliminary experience--is the watering can puncture superior? *The Journal of Urology, 186*(4 Suppl), 1700–1703 https://doi.org/10.1016/j.juro.2011.04.007.

Paye-Jaouen, A., Pistolesi, F., Botto, N., Enezian, G., Grapin-Dagorno, C., Peycelon, M., & El-Ghoneimi, A. (2015). Long-term bladder function after ureterocele decompression in children. *The Journal of Urology, 193*(5 Suppl), 1754–1759 doi:10.1016/j.juro.2014.10.115.

Peters, C.A., Mendelsohn, C. (in press). Ectopic ureter, ureterocele, and ureteral anomalies. In: A.W.P., C.A.P., Alan, J., Wein, Louis, R., Kavoussi, (Ed.), Campbell-Walsh Urology (10th ed., p. 3088). Philadelphia, PA: Elsevier Saunders.

Reed, D.C., Dodson, A.I. (1958). Ureteral injury. https://doi.org/10.1097/00006254-195806000-00056.

Romao, R. L. P., Figueroa, V., Pippi Salle, J. L., Koyle, M. A., Bägli, D. J., & Lorenzo, A. J. (2014). Laparoscopic ureteral ligation (clipping): A novel, simple procedure for pediatric urinary incontinence due to ectopic ureters associated with non-functioning upper pole renal moieties. *The Journal of Pediatric Urology, 10*(6), 1089–1094 https://doi.org/10.1016/j.jpurol.2014.04.008.

Sander, J. C., Bilgutay, A. N., Stanasel, I., Koh, C. J., Janzen, N., Gonzales, E.T., ..., & Seth, A. (2015). Outcomes of endoscopic incision for the treatment of ureterocele in children at a single institution. *The Journal of Urology, 193*(2), 662–666 https://doi.org/10.1016/j.juro.2014.08.095.

Scovell, J. M., Chan, R. C., & Khavari, R. (2014). Prolapse of a single system large ureterocele containing multiple stones in a pregnant woman. *Urology, 83*(3), e3-4. doi:10.1016/j.urology.2013.11.02.

Shah, H. N., Sodha, H., Khandkar, A. A., Kharodawala, S., Hegde, S. S., & Bansal, M. (2008). Endoscopic management of adult orthotopic ureterocele and associated calculi with holmium laser: experience with 16 patients over 4 years and review of literature. *Journal of Endourology, 22*(3), 489–496 doi:10.1089/end.2007.0312.

Shekarriz, B., Upadhyay, J., Fleming, P., Gonzalez, R., & Barthold, J. S. (1999). Long-term outcome based on the initial surgical approach to ureterocele. *The Journal of Urology, 162*(3 Pt 2), 1072–1076.

Sherman, N. D., Stock, J. A., & Hanna, M. K. (2003). Bladder dysfunction after bilateral ectopic ureterocele repair. *The Journal of Urology, 170*(5), 1975–1977 doi:10.1097/01.ju.0000085654.46865.f0.

Shimada, K., Matsumoto, F., & Matsui, F. (2007). Surgical treatment for ureterocele with special reference to lower urinary tract reconstruction. *International Jorunal of Urology, 14*(12), 1063–1067 doi:10.1111/j.1442-2042.2007.01903.

Index

Note: Page numbers followed by "f" indicate figures, "t" indicate tables.

FSC
www.fsc.org
MIX
Papier | Fördert gute Waldnutzung
FSC® C083411

Kolibri 360 GmbH
Ferdinand-Jühlke-Straße 7
99095 Erfurt, Deutschland
produktsicherheit@kolibri360.de